T0316866

Managing Coral Reefs

Managing Coral Reefs

An Ecological and Institutional Analysis of Ecosystem Services in Southeast Asia

Kelly Heber Dunning

ANTHEM PRESS

Anthem Press
An imprint of Wimbledon Publishing Company
www.anthempress.com

This edition first published in UK and USA 2018
by ANTHEM PRESS
75–76 Blackfriars Road, London SE1 8HA, UK
or PO Box 9779, London SW19 7ZG, UK
and
244 Madison Ave #116, New York, NY 10016, USA

© Kelly Heber Dunning 2018

British Library Cataloguing-in-Publication Data
A catalogue record for this book is available from the British Library.

Library of Congress Cataloging-in-Publication Data
A catalog record for this book has been requested.

ISBN-13: 978-1-78308-796-9 (Hbk)
ISBN-10: 1-78308-796-X (Hbk)

This title is also available as an e-book.

CONTENTS

FIGURES

TABLES

ACKNOWLEDGMENTS

Sincere thanks to Dr. Lawrence Susskind, Dr. Porter Hoagland, Dr. Frank Ackerman, Dr. John Ogden, and Mary and Charles E. Heber (Nanny and Popop).

ABBREVIATIONS

BAPI	Biodiversity Action Plan for Indonesia
CBD	Convention on Biological Diversity
COREMAP	Coral Reef Rehabilitation and Management Project
FAO	Food and Agriculture Organization
GDP	Gross Domestic Product
GEF	Global Environment Facility
IAD	Institutional Analysis and Development Framework
IDR	Indonesian Rupiah
MPAs	Marine Protected Areas
MYR	Malaysian Ringgit
NOAA	National Oceanic and Atmospheric Administration
NGOs	Nongovernmental Organizations
UNDP	United Nations Development Programme
UNEP	United Nations Environment Programme
USD	United States Dollars
USAID	United States Agency for International Development

Chapter 1

INTRODUCTION

1.1 Reefs and People

The hum of diesel engines on the black volcanic beach in the village of Tukad mungga signals the start of the working day in a rural yet rapidly developing village on the prosperous resort island of Bali, Indonesia. Cows can be heard in pens behind houses while the seemingly unending construction of scenic beach villas dims the agricultural noise, a sign of rapid economic transition in the village. Women wrapped in traditional sarongs and sashes leave offerings on the beach: colorful fruit and flowers in small woven baskets with incense. Men get up from their plastic seats where they were drinking sweet, white coffee and smoking clove cigarettes in order to drag the traditional *jukung* boats, brightly painted narrow vessels with slim outriggers on each side, down the beach to the water. The oldest men are wearing *udeng*—Hindu traditional headscarves—as they finish their cigarettes, carry their engines and propellers, fasten them to their boats and welcome the tourists aboard (Figure 1.1). Many tourists, both Indonesians and foreigners, line up in orange life jackets to fill the jukungs. Many are here for snorkeling and the popular dolphin-watching tours, while others are here for diving.

Although all of these men own and operate the boats for their income, nobody is undercutting his neighbor. Nobody can be seen on the beach trying to recruit last-minute visitors to fill seats on their boats. For those who have visited more developed places on the Balinese tourist route, where you cannot take two steps without being asked to buy something, this is unusual. It appears that a very organized system of boat entry, as well as fixed set prices, exists. Also remarkable is how each boatman informs his guests to not touch the corals if they are snorkeling—not in perfect English, but the message gets across. When I ask these boatmen why they don't offer discounts in order to take the customers of the boatmen in the next village over, they cite their membership in a community-based organization, with colorful Balinese names that reference their Hindu religion and the importance of the sea. They cite their common religion and the village bonds along the coast as reasons for

Figure 1.1 A Balinese boatman wearing an udeng launching his jukung.

working together to organize the dolphin and snorkel tourism, and say that the management organizations make both quality of life and social stability better. For example, one boatman says, "We live hard lives. This is why we need to work together. Angering neighbors by stealing customers is something we used to do, then over time, we learned it hurt us in the low season and we organized." Our conversation turns to the subject of the local reef, which has not fared well through the past thirty years due to dynamite fishing but is slowly coming back to life. All nearby members cite the village-based management as the reason for its return. "This is our reef, this is our living, and we care for it."

Now consider a second example, this time on a small white sandy island off the northeast coast of peninsular Malaysia. The *Perhentian Islands*, meaning "stopover" since they provided a stopover point in early shipping routes, were one of 42 peninsular Malaysian islands declared marine parks by Malaysia's 1994 *Marine Parks Order of the Fisheries Act*. The village on the smaller of the two islands has an enormous silver mosque that issues its thunderous *adzan* call to prayer five times daily, multiple food stalls where you can buy the traditional *roti canai* and many traditional-style Malay homes brightly painted in greens, turquoises and pinks to show the religious dedication of the families inside. The men have small motorboats, usually named after one of their children, parked on the beach as water taxis and ad-hoc stands where they arrange snorkeling tours for visitors at dozens of sites all over the islands with names like "Shark Point" and "Temple of the Sea." A massive new orange building sits on the beach nested behind two high-end resorts, multiple stories tall. Oddly enough, many days could go by before you could realize its purpose. The signage is in Malay, and there is minimal indication that this is a place that tourists may visit. In fact, it is supposed to be the Perhentian Islands Marine Parks Office. Due to its sheer size, I spent an entire afternoon wandering around trying to find somebody who worked there. Locals staffing the restaurants laughed, saying that the staff of the Park Office are frequently absent. They wish me luck finding someone to talk to.

One of the most striking sights, one that happened every single day, was how dozens and dozens of snorkelers—wearing life jackets, which often indicates their inability to swim—walked on the coral reef, which stands not even 10 meters out from the shore, directly in front and in clear view of the Marine Parks Office. Locals from the village sat nearby within their businesses, right next to signs in many languages that asked people not to touch or take the corals. They watched from their water taxi stands, their boats or from their restaurants. There was no sense or feeling that the visitors were damaging the reefs that form the underpinning of their livelihoods and take centuries to grow back after they are trampled. These threats to the reefs went completely unnoticed, day after day, often for the entire duration of the day, as dozens of people trampled the reef.

Why is there such a pronounced difference in the way the reefs are managed in the marine parks of Indonesia and Malaysia? Based on my initial site visits in the summer of 2013, I hypothesized that the form of governance, top-down versus bottom-up, might explain the difference. I hypothesized that bottom-up governance resulted in more successful reef management—from an ecological perspective and a socioeconomic one, whereby tangible social and economic benefits are linked to successful ecosystem management. My research asks how ecological governance affects societies, economies and ecosystems in the developing world and attempts to answer the question based on what I observed in five field sites in Indonesia and Malaysia over nine months of fieldwork between 2013 and 2015. I collected both qualitative and quantitative data, taking approximately 30 interviews and 50 surveys per site. I talked with a great many stakeholders. I took approximately 20 ecological surveys on living coral cover using timed swim methods to facilitate an integrated analysis across both social and ecological systems. I drew on institutional and socioecological systems theory to formulate my experimental design as well as my interview and survey instruments. I used a combination of qualitative thematic coding and statistical analysis to analyze my data. My findings lead to policy recommendations regarding the design of institutions for ecosystem management in developing countries.

1.1.1 Structure of the text

The structure of this text is as follows. Chapter 1 is a general overview of the importance of coral reefs, marine protected areas (MPAs) and how coastal systems such as these are managed.

Chapter 2 covers the theory, practice and policy context of coral reef management. It places coral reef management in multilateral development frameworks, discusses theories on management of natural resources, places the research in this book in the context of international development research and outlines the conventional wisdom on coral reef management.

Chapter 3 outlines the differences between the way that Indonesia and Malaysia manages their coastal resources, namely coral reefs, in a bottom-up and top-down approach, respectively.

Chapter 4 provides an overview of the case study sites across Indonesia and Malaysia used in this book. These sites include Lovina, Pemuteran, Amed, the Perhentian Islands and Tioman Island. This chapter also gives a brief overview of ecological findings.

Chapter 5 discusses integrated management of MPAs, or when MPAs are managed for both social and ecological considerations. Examples from the case study sites are given in order to show how co-managed MPAs and centrally managed MPAs have different levels of integrated management.

Chapter 6 discusses the different levels of legitimacy in co-managed and centrally managed MPAs in the case study sites.

Chapter 7 discusses the different levels of adaptive capacity present in co-managed and centrally managed MPAs in the case study sites.

Chapter 8 offers policy recommendations regarding integrated manage-ment, legitimacy and adaptive capacity of MPAs. It also offers key policy recommendations for Indonesia and Malaysia, as well as a general set of policy recommendations for coastal biodiversity conservation in general.

It should be noted that this book is based on a 2016 PhD thesis submitted to the Massachusetts Institute of Technology Department of Urban Studies and Planning for a doctorate in natural resource management planning. The research was supervised by a committee with expertise in planning, marine policy and coral biology. The technical and theoretical components of interest to an academic audience can be found in the appendices. Appendix A provides an overview of the research design. Appendix B describes data and methods. Appendix C examines the biological findings on living coral cover.

1.1.2 Societies, economies and reef ecosystems

This book focuses on coral reefs because they provide a wide range of ecosystem services and are increasingly important to social and economic well-being (Hughes et al. 2003; MEA 2003). Ecosystem services are the benefits that humans derive from ecosystem functions. These are classified according to how humans derive goods and services (Costanza et al. 1997; MEA 2003). So, they are usually classified as *production functions* that produce or provide natural resources; *regulating functions* that maintain essential life support systems; *habitat functions* that provide space for commercially valuable species; and *cultural* or *informational functions* that provide recreation, cultural values or aesthetic pleasure to humans (de Groot et al. 2002). The full range of ecosystem services provided by coral reefs are listed in Table 1.1.

Reef ecosystem services include the following: producing fish for subsistence and commercial fishing; reef tourism, which attracts people from all over the world to dive and snorkel; buffering services, which shelter communities from extreme weather and storm surge; erosion protection, which prevents the gradual loss of shoreline; and cultural and aesthetic values, whereby people value the reef for its beauty, spiritual significance and its importance as a unique natural place (Costanza et al. 1997; Peterson and Lubchenco 1997; Moberg and Folke 1999).

Ecosystem services are very important in the context of Indonesia and Malaysia. The Global Coral Reef Monitoring Network estimates that 120 million people in Southeast Asia depend directly on reefs for sustenance and to meet their economic needs. A significant proportion of the population is completely dependent on reefs for all aspects of their livelihoods (2008). Beyond those who make their living from local reefs, 60 percent of Southeast Asia's population lives on or near the coast, thus benefitting from reef regulation functions and habitat and production functions (Salvat 1992). The highest levels of coral and reef fish biodiversity in the world are in the Coral Triangle. Unfortunately, large-scale and rapid degradation threatens the ability of coral reefs worldwide, and particularly in the Coral Triangle, to provide users with ecosystem services (Hughes et al. 2005). Coral reef ecosystems pose especially difficult management challenges because high coral

Table 1.1 Reef ecosystem services as they are addressed in the published literature.

Provisioning services (Production functions)	Regulating services (Regulation functions, habitat functions)	Supporting services (Habitat functions)	Cultural services (Informational functions)
Fisheries (food)	Erosion protection	Nursery for juveniles of commercially valuable fisheries	Recreation: fishing, diving/snorkeling, general tourism
Raw materials (coral, sand)	Wave attenuation	Genetic diversity protection	Heritage values (bequest and existence)
Genetic resources	Nutrient cycling	Refuge	Traditional or religious uses
			Aesthetics
			Scientific research

and fish diversity combine with competing economic uses. This means that more users are attracted to reef-based livelihoods than the system can sustain (Christie et al. 2002).

1.1.3 Contrasting governance

The key question that drives this research is how differing modes of ecological governance impact ecosystem service delivery to stakeholders. Modes of governance are a product of the politics, policies and histories surveyed in detail in Sections 3.2 and 2.1. This book looks at two countries in the Southeast Asian region with similar socioeconomic, historical and developmental trajectories but with different approaches to ecological governance. One approach is centralized and the other is decentralized. I also refer to these as top-down and bottom-up management frameworks, respectively.

Centralized governance, characteristic of Malaysia, is the most common form of governance in both colonial and postcolonial states (Christie and White 2007; Jones et al. 2016). Centralized governance is often considered the default mode of governance. It relies on the state and its authority to command and control and is implemented by a bureaucracy (Imperial and Yandle 2005; Jones et al. 2016). In Malaysia and Indonesia, the national government holds property rights and stipulates rules regarding implementation at lower governmental tiers (Imperial and Yandle 2005). Decentralized governance has increased since the 1970s, with 80 percent of developing countries allocating some responsibility to lower tiers of government (Jones et al. 2016). Decentralization transfers responsibility for management to the populations most directly impacted by resource management decisions (De Oliveira 2002). It sees smaller groups of stakeholders instead of the bureaucracy making implementation decisions.

1.1.4 Institutions: Marine protected areas

This book compares centralized to decentralized institutions. Institutions are defined as social arrangements composed of rules, labor, financing, technologies and sanctions that determine the rate and extent of resource use (Renard 1991; White et al. 1994). The specific institutions examined in this research are MPAs, defined as coastal and marine ecosystems enclosed and reserved by law (IUCN 1988, 1994). MPAs are used to conserve biodiversity, preserve areas for tourism, restore degraded habitat or restore depleted fisheries (Christie and White 2007). Due to the global decline of coral reef ecosystems, MPAs are increasingly important in efforts to prevent or reverse degradation (Hughes et al. 2003; Mora et al. 2006). The rapid pace of the decline of coral reefs has led to pressure on governments to improve and strengthen management institutions and to expand efforts to implement MPAs (Bellwood et al. 2004). Major reasons for the decline of reefs include overfishing, pollution, disease, and climate change (Jackson et al. 2001; Harvell 2002; Pandolfi 2003; Bellwood et al. 2004).

MPAs range from those that are centrally managed to those that completely devolve responsibility and authority to community-based organizations. *Traditional MPAs*, most often studied in the Pacific Islands, are based on animist religions, social norms and taboos (Johannes 1981; Ruddle 1994). An example of this would be closing off an area for fishing based on religious beliefs, which simultaneously allow key species to recover for the next fishing season. *Community-based MPAs* see the community, most typically a village, as the primary decision maker, rule creator, monitor and sanction applier. They are implemented in places where the national government is weak, financial resources are lacking and technical capacity is limited. Such circumstances are more likely to exist in developing countries (Christie and White 2007). *Co-managed MPAs* involve power sharing between national government and communities that depend on ecosystems for their livelihoods (Pinkerton 1989; Christie and White 1997). Co-managed MPAs often grow out of community-based MPA that have matured and earned the trust of policymakers (Christie et al. 2000). Centralized MPAs, often referred to as *state* or *national parks*, are typical of MPAs in countries in the Global North. They require strong bureaucracies and legal mandates that prescribe resource management responsibilities. These often include "zoning" or deciding where the MPA is, as well as which areas can be used for different types of extraction. Also, responsibilities can include involving users in MPA design (Suman et al. 1999). Centralized management does not preclude resource user involvement (such as commercial fishermen and the dive industry), but responsibility for implementation is usually in the hands of the central bureaucracy (Suman et al. 1999). Centralized management typically has its main office headquarters in the administrative capital, with regional offices located on site.

Southeast Asia is regarded as the "global epicenter of marine biodiversity." Yet, only 12 percent of its reefs are in MPAs of any kind. National governments are rapidly creating protected areas (Mora et al. 2006; Christie and White 2007, 1047). Although some experts have tried to catalog all of the MPAs in Southeast Asia, this is a difficult and somewhat muddled task. For example, some studies place the number of MPAs (across typologies) in Indonesia at over 100 (Glover and Earle 2004); others place the estimates in the range of 40–50 (UP-MSI et al. 2002). Table 1.2 briefly describes the number, type and location of Southeast Asian MPAs.

Table 1.2 MPAs in Southeast Asia.

Location	Number of current proposed MPAs	Number of proposed MPAs	Type
Brunei	6	2+	Centralized MPAs under the Department of Fisheries under the 1984 Wildlife Protection Act
Indonesia	29	14+	Traditional management and community-based management such as "sasi" in Maluku in the *Trochus niloticus* fishery. Co-management in Sulawesi, Bali and other islands. Centralized MPAs (*Taman Nasional*) under the 1992 National Park Planning Guidelines.
Malaysia	40+	3+	Centralized MPAs (*taman negara*) under the 1972 Protection of Wildlife Act. Co-managed MPAs are in the exploratory phase in Sabah.
Philippines	180+	100+	Traditional, tribal and customary management dates back to the 1870s in some areas. Community-based and co-managed MPAs, such as the Sumilon Island MPA and the Ao Island MPA, exist in the Philippines and have a large portion of the literature dedicated to them. Centralized MPAs including the first marine park in Southeast Asia and Hundred Islands National Park off Luzon.
Singapore	2	4	Centralized MPAs including the Sungai Buloh Reserve and the southern offshore islands.
Thailand	23	0	Centralized MPAs under the National Park Act of 1961 and Fisheries Law of 1947 under the management of the Royal Forestry Department.
Vietnam	22	7	Centralized MPAs under the 1993 Law on Environmental Protection. Currently, no agency is responsible for MPA management, although the Ministry of Fisheries is beginning to take on the task. There is also no way to

Continued

Table 1.2 *(Continued)*

Location	Number of current proposed MPAs	Number of proposed MPAs	Type
			create regulations on boundaries for candidate MPA sites. Community-based MPA management does occur because of the lack of capacity from central government but is challenged by overexploitation.
Cambodia	4	1	Centralized MPAs under the 1993 Royal Decree "Creation and Designation of Protected Areas." Community-based MPAs as much of the national park system established by the Royal Decree lacks enforcement and capacity.
Myanmar	4	1	MPAs are rudimentary here, with a few centralized wildlife sanctuaries and all MPA formation laws drawn from the Forestry Department.

Source: UP-MSI et al. (2002).

Experts argue that there is a profound need to improve reef management institutions since a great many reefs continue to decline (Richmond 1993; Brown 1997; Hughes et al. 2003; Bellwood et al. 2004). Despite the consensus on the need for institutional change, there is no agreement on exactly what should change and why.

Chapter 2

THEORY, PRACTICE AND POLICY CONTEXT OF CORAL REEF MANAGEMENT

2.1 Multilateral Frameworks for Conservation in Indonesia and Malaysia

In both Indonesia and Malaysia, the turning point for ecosystem governance came in the wake of the 1992 Convention on Biological Diversity (CBD). The CBD is a global framework aimed at preventing further loss of biodiversity while encouraging sustainable use and equitable sharing of genetic resources. The CBD suggested goals that signatory states should take to prevent biodiversity loss. The *Aichi Targets* included numerical protected area goals for member nations, such as *Target 11*, to set aside 17 percent of terrestrial areas and 10 percent of coastal and marine areas, especially biodiversity hotspots, by 2020. Additionally, *Target 14* instructs signatory states to ensure that critical ecosystem services that impact livelihoods of the poor are restored or protected (Aichi Biodiversity Targets 2016).

The CBD tasks its member states with handling implementation of the treaty, and estimates that 40 million Indonesians depend on biodiverse ecosystems for subsistence (CBD 2016a). The National Development Planning Agency of Indonesia authored the *Biodiversity Action Plan for Indonesia* (BAPI) in 1994. It focused on strengthening institutions to prevent biodiversity loss using both scientific and local knowledge while ensuring a balance between ecosystem use and livelihood. A key element of the desired institutional change involved less of a focus on national parks and greater attention to devolving resource management (CBD 2016a). The key to implementing BAPI was to incorporate biodiversity conservation plans into the work of local government planning agencies. This set the stage for devolved management in the late 1990s.

Malaysia's policy for biodiversity protection can be found in its 1988 *National Policy on Biological Diversity* with 87 action plans for implementing the CBD. The policy focuses on improving centralized marine protected area (MPA) institutional frameworks, resolving confusion over federal-state and interagency cooperation and increasing the number of scientific experts working on conservation. In Malaysia, implementation of the CBD is spearheaded by the Ministry of Natural Resources and Environment, which is also the home of Malaysian marine parks (CBD 2016b).

In terms of meeting CBD obligations, Indonesia and Malaysia have several things in common. First, since both are considered "megadiverse" countries under the CBD, the formation of MPAs was deemed a top policy priority. Currently, Indonesia has 4.5 million hectares of MPAs. As of 2006, 40 percent of its coral reefs were classified as damaged, due primarily to destructive fishing (CBD 2016a). Malaysian marine parks include over 248,000 hectares of MPAs in peninsular Malaysia managed exclusively by the Department of Marine Park Malaysia. In East Malaysia, coral reef resources are likewise significant, with Sabah's 73,793 hectares of MPAs and Sarawak's 234,362 hectares of MPAs. These MPAs are managed by state agencies including Sabah Parks and Sarawak Forestry Department, respectively (CDB 2016a). Second, both implementation plans call for significant institutional changes, although which changes are required to enhance effectiveness is not clear.

Much like CBD, the *Millennium Ecosystem Assessment* recommended significant change to institutions for natural resource management. The *Millennium Ecosystem Assessment*, a global, peer-reviewed study on the world's ecosystems, found that human-caused ecological degradation has increased more rapidly and extensively in the past 50 years than in any prior period in human history. It demonstrated a need for empirical research on the specifics of desirable institutions and a gap in knowledge relating to the future directions of institutional change. The 2002 World Summit on Sustainable Development and the 2003 World Parks Congress—both multinational efforts to stem the loss of biodiversity—also called for significant institutional change and increasing the range of protected areas (Mora et al. 2006). Empirical studies of MPAs by academics note that despite their spread, weaknesses in governance and enforcement have resulted in poor performance (Christie et al. 2003).

2.2 Theorizing about Institutions and Change

The previous section summarized the growing pressures for institutional reform. The emergence of common pool resource theory has also clarified what we know about designing institutions for ecological governance. Elinor Ostrom, in her seminal work *Governing the Commons*, offered a framework based on empirical analyses of hundreds of cases, with an eye to managing for ecological functionality. This book draws on Ostrom's Institutional Analysis and Development Framework (IAD) (2009).

IAD suggests that there are strong links between eight key institutional building blocks and the health of the ecosystems they manage (Ostrom 2009; Wamukota et al. 2012). These include clear boundaries between those who are members of institutions and those who are not, rules within each institution that reflect local needs, the ability of members to change the rules, assurances from higher levels of government that institutional rules are enforceable and valid, monitoring protocols to make sure rules are followed, graduated sanctions for rule breakers, affordable dispute resolution protocols to deal with rule breakers and changes to rules and multiple scales of management from the topmost layer of government down to the local level. In addition to these eight building blocks, other scholars have also emphasized the importance of the ecological characteristics of a system and the contextual features of each community including local values, behavior and culture (Ostrom 1990; Becker and Ostrom 1995; Agrawal 2001; Ostrom 2007).

Beginning with boundaries, these must be clearly defined within any institutional arrangement. In particular, households who have the right to withdraw from or use an ecological system must be defined (Ostrom 1990). Second, rules that bind members, such as gear restrictions on fishermen, must be appropriate to local conditions. Rules can cover topics ranging from timing for extraction, gear allowances, places where extraction of resources can occur and how much of a resource can be removed. Third, stakeholder participation, or what Ostrom refers to as collective action, must empower individuals and groups to assume decision-making responsibilities (Ostrom 1990; Reed 2008). This implies that the knowledge, experiences, and preferences of local resource users should be taken into consideration along with those of government and scientific experts. The literature on stakeholder participation in resource management institutions suggests that the benefits of user involvement ought to include increased management capacity, greater potential for long-term management and a greater sense of empowerment.

Since resource users work day-to-day in the natural resource context, they may be a source of clever new options or solutions to resource management problems that may not have been previously considered by experts (Ostrom 1990; Scott 1998; Adger et al. 2005; Armitage et al. 2007; Ostrom 1990; Mulrennan et al. 2012). In many instances, such solutions may help to ensure long-term sustainability of resources by enhancing user buy-in and legitimacy (Pomeroy and Douvere 2008). While many theoretical discussions of natural resource institutions argue that participation by resource users is beneficial, there are also criticisms (Reed 2008). Participation does not occur in a political vacuum and, in many cases, power inequalities and marginalization of stakeholders still occur despite inclusion (Kothari 2001; Dunning 2015). Additionally, few empirical accounts exist linking participation to improved ecological outcomes (Renn et al. 1995; Rowe and Frewer 2000; Reed 2008).

The fourth, fifth and sixth institutional building blocks deal with rule enforcement through monitoring, sanctions and dispute resolution. Conflict is inevitable among resource users, as well as between government and resource users (Ostrom 2008). This is often caused by sharp differences in values among interested parties (Dietz et al. 2003). Dispute resolution mechanisms according to institutional theory must be low cost, locally based, readily accessible by stakeholders and capable of resolving conflicts among stakeholders and resource users (Ostrom 1990; Agrawal 2001; Anderies et al. 2004). Finally, there must be a "nested" quality to these mechanisms that allows stakeholders at different scales of management, from the national to the local, the authority they need to act and to do their jobs (Anderies et al. 2004). Designing multiple institutional tiers that work to quickly discover and resolve conflicts is critical to ensuring the longevity of institutions (Ostrom 2008).

Conflict resolution mechanisms in centralized and decentralized management institutions must deal with two related challenges: the time it takes to resolve disagreements and the perceptions of resource users. There is a need for rapid enforcement of rules and sanctions that can be handed out quickly (Anderies et al. 2004). If sanctions are not rapid, stakeholders will begin to doubt the fairness of the institutional arrangements. The endurance of institutions depends on whether stakeholders perceive them as legitimate, resource users have the information they need (so they can follow the rules and report rule breakers) and a common understanding of what the rules are (Anderies et al. 2004).

2.2.1 Socioecological systems: Comparing institutions

In addition to institutional design principles, ecological conditions must be considered. One way to take account of the integrated nature of people and natural systems is to consider institutions as complex socioecological systems, a conceptual frame that circumscribes the links between human systems (institutions and economies) and ecosystems. Socioecological systems are characterized by dynamic, nonlinear relationships where sudden shifts into undesired states can occur (Berkes and Folke 1998; Ingold 2000; Berkes et al. 2003; Armitage et al. 2007). The socioecological systems framework brings normative considerations into the analysis of institutions, and implies several non-ecological attributes of *successfully* designed resource management institutions (Armitage et al. 2007).

Scholars of socioecological systems have suggested several institutional characteristics required to ensure both social and ecological success. Successful institutions tend to incorporate *integrated management* strategies, focused on local ecosystems, economies and societies (Armitage et al. 2007); *participatory design* that is inclusive of all stakeholder groups that depend on a resource for their livelihoods (Pinkerton 1994; Olsson et al. 2004) and *adaptive problem-solving* that incorporates new learning into management decision making and allows for management changes to occur (Plummer and Fitzgibbon 2004; Berkes 2007).

What is meant by *successful*? Successful institutions for managing complex socioecological systems can be defined in two ways. Either in terms of (1) socioeconomic success, where stakeholders have been able to interact collaboratively for conservation purposes (Christie 2004), or (2) *ecological success*, where ecological balance has been maintained (Hughes et al. 2003; McClanahan et al. 2006). Sometimes, only one type of success is achieved by a management institution (Christie 2004; Brooks et al. 2006; Wamukota et al. 2012).

2.3 Significance of This Research: Development Trends and Institutional Norms

Globally in the twentieth century, natural resource management institutions grew increasingly centralized (Ostrom 2009). Colonial governments transferred responsibility for managing ecosystems from local communities to national government while post-colonial governments held onto this power (Pomeroy 1995; Christie and White 1997). Centralized governance was the most common form of governance in colonial and postcolonial states. The 1960s and 1970s ushered in a near-global consensus that national governments are responsible for environmental protection (Renard 1991; White et al. 1994; de Oliveira 2002; Christie and White 2007). Centralized natural resource governance relies on the work of experts to answer technical questions, such as where to locate MPAs in order to maximize biodiversity conservation (Alcorn 1993; Ostrom 2009).

Changes began in the 1970s with debates over which levels of government should be involved in resource management (de Oliveira 2002). Critics argued that centralized modes of ecological governance failed to prevent ecological degradation (Christie and White 1997), excluded resource-dependent communities and indigenous peoples from roles and responsibilities in resource management (Peluso 1992; Berkes 2007; Bonham et al. 2008); encouraged the overexploitation of coastal resources to satisfy European markets (Christie and White 1997); failed to take into account local knowledge and experience (Plummer and Fitzgibbon 2004) and ignored the perceptions of stakeholders when these differed from those of managers and policy makers (Peet and Watts 2004). There was substantial evidence of inefficient bureaucracies in developing countries throughout the 1950s and 1960s, failing to deliver services and often magnifying problems (de Oliveira 2002).

Since the 1980s, multilateral aid agencies and governments have promoted decentralization (Christie and White 1997; de Oliveira 2002). This devolution of power from the central government to lower levels of government has been accompanied by proponents claiming that decentralization (1) reverses ecological decline, (2) is the new norm and (3) is a pathway to reducing socio-ecological risk for impoverished, resource-dependent communities (Pinkerton 1994; Christie and White 1997; Zerner 2000; Ribot 2002; Colfer and Capistrano 2005).

2.3.1 Defining adaptive co-management

"Adaptive co-management" institutions are widely seen as an alternative to centralized management, particularly in places where centralized management has either ignored communities or worsened ecological conditions because of corruption or lack of capacity (Wells 1993; Christie and White 1997; Armitage et al. 2007). Adaptive co-management institutions combine *co-management* processes with processes of *adaptive management*. Co-management is an institutional arrangement where the national government legitimizes and supports resource user communities collaboratively managing their own resources through decision making at the local scale (Berkes et al. 1991; Pinkerton 1994; Armitage et al 2007).

Co-management institutions have both hypothesized and observed benefits including empowerment of local resource users to manage natural resources critical to their own livelihoods (Dodson 2014), observed ecological recovery of degraded ecosystems (Christie et al. 2002), lessening the need for otherwise high monitoring and enforcement costs (Sandersen and Koester 2000) and improving management efficiency and good governance (Jentoft et al. 1998). When local resource users are allowed to participate in management, it is presumed that this leads to more transparent decisions, is more likely to incorporate multiple types of knowledge and increases public trust in decisions (Lee 1994; Daniels and Walker 1996; Christie and White 1997; Richards et al. 2004). Co-managed ecosystems improve monitoring and enforcement efficiency since those who make rules and enforce them are located closer to the action and can respond rapidly (Jentoft et al. 1998).

Adaptive management encourages experimental "learning by doing," with iterative conservation interventions that can be adjusted over time based on learning (Holling 1973; Walters 1986; Gunderson 1999; Lee 1994). Designed in the 1970's by ecologists Buzz Holling and Carl Walters, who drew upon operations research and systems science, adaptive management is a structured decision-making framework that helps stakeholders use purposeful learning to make decisions (Holling 1978; Walters and Hilborn 1978; Walters 1986). The main function of adaptive management is to overcome what its creators regarded as the primary obstacles to decision making under uncertainty (Gunderson 2001). Increasingly, governments are promoting the integration of adaptive management into the way they manage their environments, either through policy design or the use of regulations (Camacho et al. 2010).

2.3.2 Adaptive capacity

Adaptive co-management institutions are thought to address a challenge that top-down administration can not: building *adaptive capacity* in a socioecological system. Adaptive capacity is the system-wide ability to respond to unanticipated disturbances while maintaining key ecological, economic, and social functions (Armitage 2005). Adaptive capacity is the product of both learning at the institutional level as well as the ability to employ "innovative solutions in complex social and ecological circumstances" (Walker et al. 2002; Folke et al. 2003; Armitage 2005, 706). Institutions characterized by adaptive capacity enable stakeholders to engage in learning with the "flexibility to experiment, and adopt novel solutions [...]" (Olsson et al. 2004). Adaptive capacity is itself an innovative solution(s) that communities can adopt in order to respond to crises, or what Adger refers to as the "expansion of the range of variability" with which an institution may cope (Adger 2003, 32). The literature on adaptive capacity breaks the concept into four parts: uncertainty and ultimate collapse within a socioecological system; biodiverse ecosystems and redundant ecosystem functions managed by institutions with similar diversity and redundancy; equal prioritization of local and scientific knowledge; and culminating in the opportunity for stakeholders, ranging from local communities to government officials, to react in "self-organized" and innovative ways for sustainability (Folke et al. 2003).

Multilateral aid agencies such as the World Bank, the Food and Agriculture Organization of the United Nations (FAO), the United States Agency for International Development (USAID) and the GEF, as well as NGOs such as World Wildlife Fund and Conservation International have spent millions of dollars trying to build capacity in countries seeking to implement adaptive co-management programs (see Jul-Larsen et al. 2003; USAID/COMFISH 2016; Comoros 2015; CTI 2014). Before the 1990s, the World Bank rarely supported programs that allowed for the participation of local resource user stakeholders. More recently, however, it has changed its approach for three reasons: technical, nonparticipatory approaches have lost favor; governments face fiscal problems requiring partnerships with local stakeholders where they fund their own efforts; and increasing faith in the ability of local stakeholders to perform management tasks (Uphoff 1991; Christie and White 1997).

2.3.3 Criticisms of adaptive co-management

Given the financial pivot in the multilateral aid sector towards co-management, it is important to note that co-management is not without its critics. Co-management has been criticized for proceeding without systematic, comparative and empirical work linking management frameworks to biophysical outcomes (Cinner et al. 2012; Wamukota et al. 2012; Stevenson and Tissot 2014). Empirical work that includes both quantitative and qualitative data, drawing on the experiences of stakeholder groups, is also scarce (Cinner et al. 2012; Wamukota et al. 2012; Stevenson and Tissot 2014). Studies suggest that scholarship on co-managed systems, especially co-managed reefs, claims an underdeveloped causal link between institutions and reef ecosystem outcomes, as well as a profound weakness in demonstrating that all of the hypothesized benefits of adaptive co-management systems actually occur in practice (Stevenson and Tissot 2014).

A second line of criticism argues that co-management perpetuates preexisting inequalities and social relations in which well-connected locals engage in the same types of corruption, exploitation and rent-seeking that previous (often corrupt) central authorities engaged in (Ostrom 2009). Instead of increasing local stakeholder participation, decentralization may simply allow local-scale elites to control resources, in effect creating local autocracies (de Oliveira 2002). As mentioned in Section 3.2, there is evidence that Indonesian decentralization has resulted in corruption similar to that which existed during the Suharto-era dictatorship, but with a new set of actors in control (Hadiz 2004). Thus, more decentralization cannot be presumed to be desirable. Instead, *democratic* decentralization should be the goal Hadiz suggests that co-management under the Indonesian brand of democracy may fall short of, as it has been co-opted by a network of well-placed individuals across the public, private and military sectors that have used it to their advantage (2004). Decentralization has also been criticized for allowing primarily Western environmental NGOs (ENGOs) to further their own interests, while stereotyping the developing world as a homogeneous entity and seizing the "voice" of the community in order to further other organizational goals (Peet and Watts 2004). This line of criticism hinges on the problematic use of the term community. Some studies argue it is a fundamentally misleading term that erroneously suggests communities are homogeneous, when in fact they are dependent on the same long-standing power networks that continue to haunt Indonesia (Berkes 2011). The term community may also mask internal complexities, such as socioeconomic differences between well-connected elites and low-income inhabitants.

Institutional analysis is one way to address such criticisms. It allowed this research to examine rule-making patterns and enforcement mechanisms in order to sidestep the problem of viewing the community as a homogenous entity (Ostrom 1990). In other words, instead of looking at a village as the unit of analysis, this book uses the institution. Institutions are the primary unit of analysis.

2.4 Conventional Wisdom on Reef Management

This section presents the conventional wisdom on institutions for reef management in developing countries in the tropics. There are five major themes in the scientific literature on reef management. These are (1) whether certain institutional design principles are effective, (2) how stakeholder perceptions vary across institutions, (3) linking institutions to ecological outcomes, (4) linking institutional success to contextual factors such as demography or proximity to markets, and (5) defining "successful" management institutions.

2.4.1 Designing institutions for reef management

Many case studies on reef management discuss Ostrom's design principles for institutions, with a focus on rules and enforcement. In *Governing the Commons*, her second design principle assumes that rules can restrict time, place and gear typology but that they must be appropriate to local conditions (1990). Published studies show that fewer rules lead to enhanced compliance among those who use the reef for fishing. For example, Cinner and Huchery (2014) compared reef management in East Africa and Indonesia in terms of the number and configurations of rules, finding that when there are more rules, there is a negative relationship with compliance. They also found that depending on the type of institution involved, perceptions of the benefits of the rules to stakeholder livelihoods varied, with traditional regimes having the highest levels of compliance and co-managed regimes following at a close second. In places with management institutions rooted in community traditions, such as religious restrictions on fishing, rules do not need to limit access altogether but can simply regulate gear. For example, McClanahan et al. (2006) demonstrated that allowing resource users open fishing access while limiting gear typologies to a line and pole did not lead to much damage to the reef. To summarize, places with fewer, simpler rules grounded in local traditions tend to yield more successful reef management.

Case studies also focus on Ostrom's third design principle on collective choice arrangements. Ostrom argues that people who are affected by the rules should be able to participate in making them (Ostrom 1990). Inadequate participation opportunities for resource users have been shown to limit the effectiveness of reef management. Sanderson and Koester 2000 examine a co-managed reef in St. Lucia. They found that government claims that local users were participating in reef management were not true. In this case, a well-meaning NGO along with a government agency made all of the management decisions, while framing the process as a "co-managed process." This study also shows the difficulties surrounding the mindset change required in management agencies to transition to a co-managed institution. Because officials were used to regulating the access of resource users, it was difficult for them to begin to include them in the decision-making process.

Ostrom's sixth design principle argues that rapid access to dispute resolution mechanisms is necessary for the survival of reef management institution and surrounding ecosystems. Sanderson and Koester's case study of St. Lucia shows that negative stakeholder perceptions of the fairness of conflict resolution mechanisms limit effective management of reef systems (2000). Sanderson and Koester also showed that resource users felt that conflict resolution processes were designed in an unfair, selective and biased manner. This illuminates a major pitfall in reef management and, more broadly, in co-management: the propensity to view the community as a homogenous entity when it is not. In this case, both the rules and the dispute resolution processes appeared to favor fishermen who used one type of gear and excluded those who preferred to use another gear type. Since the co-management process defined not just the community but also the fishermen as one group, important stakeholders had no influence on the way resources were managed. The authors suggest that in forming co-management processes, stakeholder groups should never be defined by outsiders, because significant differences in the way stakeholders use a resource, the way they wield power and the way they seek to advance their self-interest will often be hidden from view. Additional divisions among stakeholder groups include the way the tourist industry and its clients (mostly divers) are a relatively new user group compared to fishermen, and typically differ from fishermen along racial and class lines. The authors showed how their language skills, wealth and social experiences gave them an advantage in the reef management process compared to fishermen. In sum, institutions that have an accurate knowledge of stakeholder interests, take account of preexisting power structures and genuinely include resource users are better able to endure.

Ostrom's eighth principle for institutional design says that institutions should be "nested enterprises" in which conflict resolution, monitoring and enforcement are organized within multiple scales of governance, from the national government down to the local scale (Ostrom 1990). The literature on reef management demonstrates that although stakeholders perceive co-managed systems as benefitting communities and reef ecosystems, without links between local institutions and higher levels of government, management effectiveness will suffer. Cinner et al. (2009b) examined co-managed coral reef frameworks in Kenya and Madagascar. They found that a lack of nested institutions for resolving conflicts led to delays on resource users who block management efforts, even though conflict resolution mechanisms were available locally. Thus, nested and rapid dispute resolution protocols are key to enduring institutions. Note that so far, most studies that focus on Ostrom's design principles measure the success of the institution by (1) resource user compliance or (2) the ability of the institution to endure.

2.4.2 Reef management and stakeholder perception

Beyond Ostrom's institutional design principles, other studies have focused on the variability of stakeholder perceptions ranging from positive to negative across institutional types. It has been shown that co-managed institutions tend to have stakeholders that their livelihood benefits from management, compared to centrally administered national parks. Cinner and Huchery (2014), for instance, compare three institutional types: co-managed regimes, co-managed national park hybrids and customary or traditional management regimes. They look at the relative levels of stakeholder compliance and analyze stakeholder perceptions of their own livelihoods. Stakeholders in traditional management institutions grounded in local customs perceived livelihood benefits from membership whereas the other two groups did not.

2.4.3 Linking ecological outcomes to institutions

One of the key criticisms of institutional theory as it applies to reef management policy is that too few studies link institutions to ecological outcomes (Cinner et al. 2012; Wamukota et al. 2012; Stevenson and Tissot 2014). Christie (2004) reviews four case studies of ecological functioning in terms of reef fish biomass, target species abundance, and coral cover, along with institutional design components of participation and dispute resolution. They find observable changes in improved ecological outcomes following the creation of a new management institution, even where dispute resolution mechanisms fall flat in the face of preexisting community conflicts. Thus, resource users in cases still had disagreements with one another but were able to follow the rules of the MPA, resulting in improved coral health.

There is a noticeable tradeoff in the level of detail paid to socioeconomic characteristics of case sites in studies that have a wealth of ecological data. Socioeconomic variables are rough-grained in this study, while ecological variables are fine-grained.

Other studies attempt to link the status of reef ecology to a range of decentralized or centralized management institutions. For example, traditional institutions governing reef management with longstanding presence in the community (such as religious beliefs about when and how to limit catch) can lead to healthy ecosystems, probably due to the in-depth local understandings of human-environment relationships. McClanahan et al. (2006) compare the impact on reef ecology of institutional typologies that range from national parks, to traditional management, to co-management, while also examining contextual factors such as proximity of case sites to markets, economic data, and population densities. Along the same lines, Cinner et al. (2009a) examines the relationship between economic development, human population density and specific ecological indicators of reef health, such as the complexity of its structure (termed its *rugosity*) and reef fish biomass. They link several demographic and economic factors to diminished reef ecological functionality, suggesting that some social and economic patterns have demonstrable links to ecological outcomes.

In studies that link institutions and ecological outcomes, there is a broader debate over whether key differences across management institutions can be tied to ecological outcomes such as fish biomass, coral cover, or fish abundance. Christie et al. (2002) show empirical differences in living coral cover before and after a community-based MPA was established in Balicasag and Pamilacan in the Philippines. They note that complex causes hinder a clear reading of their findings, because bleaching as a result of elevated sea-surface temperatures in combination with crown-of-thorns starfish outbreaks also led to a reduction in living coral cover. It has also been shown that fish biomass and size have strong responses to different management institutions, whereas other ecological variables such as coral cover, target species density and overall abundance of fish do not (McClanahan et al. 2006). It is important to note that these responses are at different scales of natural or man-made disturbances, with coral cover primarily impacted by local, regional and global stressors. The authors offer several possible reasons why ecological variables did not vary across management typologies: management could be ineffective in one or all of the cases, sampling was not powerful enough, local fishing pressure was not severe enough to warrant management closures or, generally, ecological parameters do not respond as strongly as previously thought to management interventions. Valuable, target species have also been demonstrated to respond to changes in the structure of management institutions. Hamilton et al. (2011)

examined the relationship between community-based institutions and valuable grouper species with aggregate spawning behaviors in Papua New Guinea. They found that community-based institutions led to population recovery in fish that aggregate to spawn.

Cinner et al. (2012) attempt to link co-management institutions at 42 sites in the Indian Ocean with ecological outcomes. They failed to find any correlation between institutional design and ecological outcomes, suggesting that in some cases, they may have no substantive effect on ecological outcomes. Their findings could also be due to some of the largest problems with the literature on management institutions and coral reefs: a lack of data on ecological trends before and after the creation of new management institutions, with controlled comparisons to non-managed sites, as well as issues with endogeneity whereby institutions cannot be randomly assigned in real life, and may be just one cause of ecological functioning. Several studies do use control sites, with managed reefs compared to non-managed reefs. For example, McClanahan et al. (2006) present findings that discuss links between management institutions and ecological functioning, and use their control sites to tie biomass to institutional design and, by contrast, rule out coral cover as having links to institutional arrangements. The broader takeaway from all of this research is that there are remarkably few studies that show a correlation between institutional arrangements and ecological variables, and there is no consensus yet on how best to assess these connections.

Several studies examine *contextual factors* that may impact institutional longevity and ecological health (Agrawal 2001). These include the roles that nearby markets, economic sectors and population densities might play in incentivizing compliance or lack of compliance. It has been shown, for example, that in places with fewer markets and more bartering, lower human populations and higher per capita income, healthier reefs can be expected (McClanahan et al. 2006). Some studies have shown that in places where there are expensive conservation programs created by NGOs, multiple and diverse types of income per household and higher levels of education, one could not necessarily find healthy reefs (McClanahan et al. 2006). This suggests that institutions and contextual factors are somewhat difficult to link to ecological health.

Other case studies of co-managed reefs have discussed ways to measure the effectiveness of different reef management institutions. Both Wamukota et al. (2012) and Christie (2004), show that demonstrating relationships between ecological variables and institutions is not the only way to evaluate management success. Instead, they show evidence of shifting stakeholder attitudes and behavior change regarding rule compliance or collaboration. Other types of successes include improvements in standards of living, enhanced equity, trust and social capital (Pollnac and Pomeroy 2005).

Chapter 3

GOVERNING NATURAL RESOURCES IN INDONESIA AND MALAYSIA

3.1 Indonesia's Road to Decentralization

Indonesia is currently expanding its network of co-managed marine protected areas (Mpas) to supplement its preexisting network of centrally managed marine parks. This decentralized approach to ecological management is a product of the country's colonial and postcolonial political history. In the last 30 years, Indonesia has devolved significant responsibilities to local communities while continuing to maintain national parks.[1]

Although they are not the focus of this book, some attention must be paid to the centrally managed Mpas in Indonesian waters that are known worldwide by divers and coral reef experts for their high levels of biodiversity and coral cover. In recent years due to their popularity, the agency responsible for tourism in Indonesia has begun to market these parks as the "Magnificent Seven," and they include The Thousand Islands, Karimunjawa, Wakatobi, Taka Bonerate, Bunaken, Togean and Cendrawasih.

---

1 While Indonesia is home to some of the most biodiverse and popular national parks in the world for divers, this book focuses on the co-managed Mpas of Indonesia for several reasons. First, the author was unable to obtain proper permitting for fieldwork in the national park boundaries, which can be cumbersome even in the best situations due to high levels of corruption, bureaucratic inaction and Indonesia's ever-changing immigration laws. Second, the costs of running surveys in national parks versus co-managed parks were too high given the budget of the research and the permitting process mentioned above. Last, the design of the research used nation states and one of their modes of governance (centralized or decentralized) as the unit of analysis. Were this to be expanded to compare modes of governance within countries as well as across countries, the time and funding required for this research would have needed to increase significantly. Thus, although the lack of survey sites in Indonesian marine parks is a flaw to the research, the ultimate aim is to compare co-managed sites to centrally managed sites in two countries in Southeast Asia, and that goal was ultimately met.

3.1.1 Colonial legacies and changing governance

The story of how natural resources are managed in Indonesia begin in precolonial times. However, the very idea of a unified Indonesian government is actually a recent one, with the notion of a unified nation state first appearing under a nationalist movement that began in the 1920s.

Prior to the Dutch colonial era, Indonesian natural resource management focused on extraction and it varied from island to island. For example, early Indonesian kingdoms such as Mataram in Java and Srivijaya in modern-day Sumatra had a highly centralized natural resource management policy during the "Golden Era" of Indonesian history spanning 1–1500 CE. Rulers oversaw strict hierarchical, feudalist-style systems that directed the peasantry to conduct forest harvests without limits (Brown 2003; Bourchier and Hadiz 2014).

The dominance of these decentralized kingdoms, or the early kingdoms described above, continued until the Dutch colonial era. Dutch colonization occurred through its East India Company (VOC) for the purpose of extracting, exporting and holding a monopoly over Indonesian natural resources including spices, minerals and mining resources. The colonial era began in the early 1600s with the establishment of modern-day Jakarta, known then as Batavia, in 1610 (Thorburn 2002). The Dutch were able to centralize governance effectively on Java by the 1830s. But given the large extent of the "East Indies" as Indonesia was known at the time, control was difficult and expensive to establish and maintain. Failures to centralize rule over islands such as Bali and Borneo persisted until the 1900s.

James Scott, in *Seeing Like a State*, describes the economic motive for Dutch centralization as one in which commercial logic and bureaucratic logic merged in an effort to maximize the investments of the colonial power (1998). Increased centralization was critical to the success of the Dutch export-oriented business model, selling spices to European markets at 17 times their value due to the Dutch monopoly (Brown 2003; Bourchier and Hadiz 2014). Additionally, Dutch colonial governors heavily taxed local landowners and implemented a strict feudal system of forced cultivation. Between the 1830s and 1870s, this led to a mass repatriation of much of Indonesia's wealth (de Vries 2011). Centralization led to dozens of bloody wars and widespread, brutal oppression of indigenous leaders throughout the 1800s and the early 1900s. After World War II, Indonesians led by Sukarno won their independence from the Dutch (Brown 2003).

3.1.2 New Order Indonesia and centralized control

The effort to centralize state control continued after colonial independence. Sukarno proclaimed Indonesian independence in 1945, beginning a long struggle to unite the island under one government. Indonesians rarely viewed themselves as Indonesians and, instead, viewed their identity as linked to their island; even today it is common for people to introduce themselves as *Javanese* instead of *Indonesian*. The difficulties in controlling the new state of Indonesia led to many ethnic tensions, as well as an early political legacy of dictatorship. Colin Brown, an Indonesian historian, notes that many of these struggles have continued into contemporary times, such as those involving East Timor and Papuan separatists (2003).

In 1965, in response to the challenges of uniting Indonesia, the authoritarian New Order policies of the Suharto dictatorship created centralized governance, weakening local, traditional, village-based and indigenous management frameworks that managed to survive the Dutch colonial era (Thorburn 2002). Postindependence Indonesia centralized its resource governance. The state forestry service, for example, was a direct legacy from the colonial era and was tasked with protecting valuable species such as teak (Peluso 1992). In colonial Indonesia and in Suharto's Indonesia, state-owned land was used to grow teak trees owned by the state (Henley 2008).

The devolution process began with the fall of Suharto's 32-year dictatorship. He resigned in 1998 following the social and economic shocks of the Asian fiscal crisis of 1997. The shocks and the subsequent social unrest led to his administration's crackdown on political opponents. This was followed by insurmountable protests over what many saw as highly oppressive leadership. Beginning in the late 1990s, Indonesia underwent major political changes, termed the *Reformasi* or "reforms" in Indonesian. Many of the functions of the previously highly centralized government were devolved to the district, provincial or municipal levels (Thorburn 2002).

3.1.3 Contemporary rise of co-management

The two most important laws resulting in decentralized MPAs in Indonesia were *Law 22 on Regional Government* and *Law 25 on Fiscal Balance between the Center and the Regions* (2002). These laws delegated powers and finances away from central government in Jakarta to the district and municipal governments. The implications for natural resource management were that local governments could create a spatial planning policy, determine ecosystem use, devolve ecosystem management to villages and manage budgets to these ends. In terms of MPAs, local governments such as *Kabupaten* (known as regency governments in the Indonesian system) and *Kotas* (city governments) were given power over coastal and marine ecosystems within four miles of the shoreline and were permitted to devolve this power to villages. Decentralized resource management is supported through various multilateral development schemes.

For example, the Indonesian government, since the fall of Suharto and the subsequent *Reformasi* in 1998, has worked with the World Bank to finance a three-phase program, partially funded by the Global Environment Facility (GEF), called the Coral Reef Rehabilitation and Management Program (COREMAP 2007) (CTI 2014). Centrally managed MPAs do exist in Indonesia, but additional priority areas characterized by high levels of biodiversity are the focus of expanded efforts to devolve management (World Bank 2015c). This program assists the government in devolving management to the lowest scales of villages or *desa*. The reason behind the transformation was the fall of the Suharto regime in 1998, and with it, the implementation of the *Otonomi Daerah* Act No. 22/1999 ("Regional Autonomy Act") that transferred power from the central to the local government. This legislation was the end of a struggle between local governments for control over their resources against the centralized dictatorship of Suharto. The contemporary reforms are taking an explicitly co-managed approach, defined as a management institution with the following characteristics: participation by local and diverse stakeholders, integration of reef management choices with the broader socioeconomic needs of a community, empowerment of the community to act as resource managers and employing local environmental perceptions to make management decisions (White et al. 1994; Christie and White 1997; Christie and White 2007; Cinner et al. 2009b; Kittinger 2013).

The goals of capacity-building in local communities include empowering districts and community-based institutions, increasing income, increasing capacity to manage ecosystems, establishing sustainable fishing practices, public awareness campaigns and educational programs (World Bank 2015c). Despite these goals, some scholars of Indonesian politics criticize the failure of decentralization, which has been unable to completely root out old forms of corruption. Thus, some describe a new set of "predatory networks of patronage" that lessen local empowerment (Hadiz 2004, 3). Some case studies have shown that even though a co-management system was in place, local elites were in a position to co-opt the system at the village scale (Agrawal and Gibson 1999).

The spread of co-management still faces many obstacles, as evidenced in the context of the creation of several large, centrally managed MPAs (Henley 2008). For example, when the government tried to establish a national park in Bunaken, Sulawesi, in 1991, it searched for examples of community-based management practices that might be incorporated into the MPA plan. Because the coral reefs were large and pristine and the local population was rather small, there had never been evidence of resource scarcity. Thus, it was hard to justify management interventions. Moreover, if a part of the reef became degraded, there was always additional reef to move onto (Merrill 1998; Henley

2008). Similarly, when the Wakatobi MPA was created in 1996, the government did not include the community viewpoint in any zoning considerations and the park's creation excluded many locals from livelihoods held for generations (Elliot et al. 2001). Couple the historical lack of community involvement in the creation of centrally managed MPAs with the well-documented shortcomings of Indonesian resource governance, including the lack of funds, few technical experts and corruption, and it is clear why the spread of co-management institutions faced many challenges (Pet-Soede et al. 1999).

3.2 The Origins of Centralized Malaysian Governance

In contrast to Indonesia, centralized governance characterizes the Malaysian context and has since the state was created in 1957 (Pomeroy 1995). Malaysia and Indonesia have entwined histories. For much of their early history, Indonesian and Malaysian territories were blurred, such as in the case of Srivijaya (modern Sumatra), considered by historians to be Malay as well as an Indonesian territory. There was a great deal of contact between Javanese, Balinese, Sumatran and Malaysian societies. For example, all Indonesian goods passed through the Straits of Melaka before continuing on to the Middle East and Europe. Both Malaysia and Indonesia share common cultures, adopting Islam in the peak of trade to better infiltrate trade networks of the wealthy Middle East. The first real instance of borders between Indonesia and Malaysia was in 1824 with the Anglo-Dutch treaty, which officially separated British Malaysian colonial territories from Dutch Indonesian territories (Matheson-Hooker 2003).

3.2.1 Precolonial kingdoms

Early legacies of centralized Malaysian governance can be found in the example of the sultanates of Melaka and later in Johor and Kedah around the fifteenth and sixteenth centuries. Primary schools teach young Malaysians today that these early regimes, characterized by unwavering loyalty to the sultan, constitute the Malaysian ideal of good governance. A strictly hierarchical way of life prevailed. As in the Indonesian kingdoms of the same period, the elites ruled and peasants pursued agricultural, mineral and forestry extraction (Matheson-Hooker 2003).

3.2.2 British colonial rule and centralization

In the British colonial era, in order to maintain its economic interests in Malaysia, Britain installed its own advisors, termed "Residents," who became the actual rulers of the territory (Matheson-Hooker 2003). By 1895, the British had begun to centralize governance by combining four sultanates as the Federated Malay States with one Resident with centralized authority based in Kuala Lumpur (Andaya and Andaya 2001). The remaining Unfederated Malay States ultimately fell under centralized colonial rule by 1914. Following World War II, the Unfederated and Federated States were merged by the British to form the Malayan Union with a British colonial governor at its head. Although real power had been lost during the formation of the Federated and Unfederated States, the sultans formally signed over power to the colonial governor. This created social unrest that ultimately led to independence in 1957.

The British colonial government controlled all natural resources in coastal areas. This centralized management structure oversaw fishing exploitation, gear types and methods, and it regulated export to European markets (Christie and White 1997). Early reports from Malaysia (Malaya) by British colonial administrators noted that community-based fisheries management occurred in coastal communities. Malay *Tomlay* people were not allowed to "take beyond a certain amount per day" even when fish were plentiful, meaning that communities were setting their own quotas on their coastal resources (Christie and White 1997, 158). Once colonial rule did away with these local quotas, the result was the over-exploitation of Malaysian fish stocks (Christie and White 1997).

3.2.3 Contemporary Malaysia

Contemporary Malaysian governance is a parliamentary bureaucracy with a chief executive, or prime minister, in charge of policymaking. Malaysian policymaking can be best described as a top-down "bureaucratic act emanating from the Prime Minister's Department" (Khai Leong 1992, 205). In order to implement policy, a large bureaucracy supports the prime minister's office to do so. Malays dominate this bureaucracy although they are only 58 percent of the population (with approximately 30 percent Chinese and 10 percent Indian). The bureaucracy is where natural resource management policy is determined and then, in a few instances, opened to the public for limited discussion.

Centralized governance in the contemporary Malaysian federal system is outlined in the 1957 Malaysian constitution; federal agencies make policy while state and local authorities implement it. The federal government is responsible for making policy in the name of the national interest. This includes the law authorizing the formation of MPAs. State governments play a supplementary role, in this case managing the land on the islands closest to the MPAs and in the marine parks. Local authorities, such as district and land offices, implement the state government's decisions. Local authorities are also responsible for physical development and land-use planning on islands that fall within the marine parks (ICRI 2010). Much like Indonesia, Malaysian historical resource governance is dominated by policies that focus on commodities, primarily oil palm products. Beginning in the 1990s, both Indonesia and Malaysia added policies aimed at biodiversity conservation (Yusof and Bhattasali 2008).

3.2.4 Malaysian Marine Parks

The Department of Marine Parks Malaysia manages coral reefs. The former prime minister Dato' Seri Dr. Mahathir Mohamed directed the Department of Fisheries to form the Malaysian Marine Parks, first declaring the islands off of the state of Terengganu a marine park in 1983. The Malaysian MPAs banned fishing on reefs included in the parks. Unfortunately, this area was home to fishing villages, which saw their way of life collapse almost overnight. In addition, there is no published information on the fishing villages that disappeared with the creation period of Malaysian Marine Parks. Two years later in 1985, government declared that fishing was prohibited in any ecosystem falling within three kilometers from the shorelines of the 22 islands in Peninsular Malaysia. A financial mechanism and a bureaucratic home for Marine Parks was established in 1993 with the Marine Park and Reserve Trust Fund and the National Advisory Council for Marine Parks and Marine Reserves in the Ministry of Agriculture.

The constitutional powers of the Malaysian government to establish MPAs can be found in Schedule 9 of the Constitution. This grants the federal government power over marine and estuarine fishing. Since management powers are still classified as fishing-related, the law under which MPAs fall is titled the Fisheries Act of 1985 (Saad 2013). Coral reefs are a grey area under this legislation. They are located on the seabed, so states can claim them, but since living resources are defined legislatively as "fish," they are commonly defined as federal subjects (Saad 2013).

The year 1994 is generally regarded as the year when Marine Parks took their current form, with local offices on the islands, management plans, as well as support from partner nongovernmental organizations (NGOs). Additionally, the 23 islands (including those in the states of Terengganu and Pahang included in this study) saw fishing bans at two kilometers from their shorelines, along with no-take rules for corals and other marine organisms. Another bureaucratic reshuffling occurred in 2004, when the Marine Parks division was moved from the Department of Fisheries to the Ministry of

Natural Resources and Environment. With this reshuffling came an increased focus on *managing* coral resources as opposed to enforcing simple fishing bans. By 2015, Malaysia proclaimed itself the leading Southeast Asian nation in biodiversity conservation as measured by its success in marine park management. The Department of Marine Parks Malaysia became its own individual department in 2006, employing approximately 200 people (Laman Web 2016). Additional stated objectives of the Department of Marine Parks Malaysia are to protect biodiversity, conserve endangered species habitat, establish management zones, deal with recreational use and carrying capacity and build public awareness. Everyday, Marine Parks officers are supposed to offer education programs, maintain public facilities, provide technical advisory services related to the ecosystem, enforce the fishing ban and no-take laws, oversee conservation programs, construct artificial reefs, undertake research and issue permits for third-party groups such as NGOs and universities to do research (Laman Web 2016).

Decentralized governance cannot be found in Peninsular Malaysia. Several case studies of Malaysian resource management note that among Malaysian fishermen, notions of self-governance or cultural attitudes that could form the bases for creating local management institutions are absent. Most importantly, studies have found that fishermen did not want self-management, preferring a centralized approach (Yahaya and Yamamoto 1988; Pomeroy 1995).

Malaysia is in the early stages of allowing more community participation in the way it manages its Marine Parks. In 2007, the federal government made an effort to modify its top-down approach to governance by seeking financing and expertise from the United Nations Development Programme (UNDP) in an initiative called *Conserving Marine Biodiversity Through Enhanced Marine Park Management and Inclusive Island Development* (UNDP 2013). This effort engaged villages that once made their entire living off fishing, including Redang and Tioman. Management plans were drafted for the islands through what the UNDP calls a "consultative process" (UNDP 2013). The new management plans prohibited fishing within two nautical miles of the coast, while large vessels needed to be more than five miles from the shore. UNDP also retrained 484 local people who previously worked as fishermen to become boat guides and tourism industry workers as a means of easing the economic transition for those left jobless after MPA creation. It remains to be seen whether similar programs will gain traction in Malaysia.

In the five Marine Parks in peninsular Malaysia, two key things make governance difficult. First, there is a legal mismatch that makes sanctions difficult to enforce. The original legislation that set up Malaysian Marine Parks is now outdated since it falls under the Fisheries Act. That Act gave enforcement power to the Ministry of Agriculture, even though Marine

Parks were subsequently moved to the Ministry of Natural Resources. Also, since state governments have jurisdiction over land (including the islands) and resources three miles offshore, there is some overlap and confusion regarding who exactly is responsible for which aspects of enforcement. The Fisheries Act also set up a federal-level multiagency body to advise the Department of Fisheries on Marine Parks, but this has not resulted in clarifications regarding responsibilities for governance.

In sum, the Indonesian context has for nearly 20 years devolved management authority to the local scale, but remains troubled by preexisting power dynamics, including networks that date back to Suharto's New Order dictatorship. Malaysia, on the other hand, has deeply centralized governance. All major decisions come from Kuala Lumpur and are implemented by state and local governments. There are local offices of the national Department of Marine Parks. Malaysian bureaucracy creates difficulties for stakeholders caught up in overlapping jurisdictions and confused allocation of responsibilities. Both Indonesia and Malaysia face serious degradation of their terrestrial and marine environments following legacies of exploitative colonial rule and development-oriented resource management in the postindependence years. Several multilateral meetings in recent years have resulted in significant pressure for institutional reform, primarily focused on creating, improving and expanding protected areas. The following chapter discusses these frameworks.

Chapter 4

CASE STUDY SITES AND
THE CORAL TRIANGLE

4.1 Situating This Research in Global Environmental Research Agendas

This chapter situates this book in the wider global development context. The 1992 Rio meetings, generally viewed as a turning point in the history of sustainable development, named research priorities for linked human and environmental systems in the context of environmental degradation. These priorities included (1) analyzing the social determinants of environmental change, (2) understanding institutions for natural resource management and their effects on biophysical systems and (3) improving governance and societal response (Ostrom 1990; NRC 1999; Moran 2010). These three research priorities must be analyzed comparatively across regions and ecosystems, and in a way that is spatially explicit, anchored in a context (Moran and Ostrom 2005). Section 2.1 describes urgent calls for research on institutional change in the *Millennium Ecosystem Assessment*, the individual country plans of the CBD, the proceedings of the World Summit on Sustainable Development and the World Parks Congress. I draw on these combined calls for research on institutional change by comparing marine protected areas (MPAs) in Malaysia and Indonesia for ecological and social outcomes using a socioecological systems framework.

4.2 Why MPAs in the Coral Triangle?

This book compares Malaysian and Indonesian reef management institutions, specifically MPAs, for several reasons. First, MPAs are the main social arrangements that people use to manage coral reefs. Second, Malaysia has centralized reef management through its federally managed marine parks, whereas Indonesia is expanding co-managed MPA coverage (while still possessing a range of centrally managed MPAs). Third, as outlined in Section 2.1, the CBD and the corresponding national plans for Indonesia and Malaysia

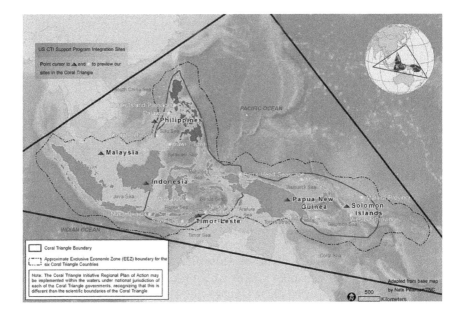

Figure 4.1 The Coral Triangle. *Source*: The Coral Triangle Initiative on Coral Reefs, Fisheries, and Food Security (CTI-CFF). http://www.uscti.org/.

on biodiversity conservation place MPAs at the center of their governance strategies for protecting coral reefs. The CBD, the 2001 World Summit on Sustainable Development and the 2003 World Parks Congress called on governments, stakeholder communities and nongovernmental organizations (NGOs) to create a global network of MPAs (Vernon et al. 2009) and thus they are increasingly popular conservation tools.

This book looks at five field sites in the Coral Triangle (Figure 4.1) for three reasons: because of the global status of the Coral Triangle as a biodiversity hotspot, the link between at-risk communities living in poverty and ecological health in the Coral Triangle Region and the large economic value of Coral Triangle reefs. The biodiversity of the Coral Triangle is unlike anywhere else on the planet. Indonesia, for example, has 77 percent of the world's coral reef species, over 500 species of corals and half of the world's reef fish species (Glover and Earle 2004). The Coral Triangle region overall has 76 percent of the world's species of coral and 52 percent of the world's fish species making it a priority for global biodiversity conservation (Veron et al. 2009). The Coral Triangle is a critical conservation concern, since it includes six developing countries within which millions of impoverished people depend on healthy reef ecosystems for livelihoods ranging from fishing to hospitality (Salvat 1992; Moberg and Folke 1999; Sheppard et al. 2009). Studies have attempted to quantify the economic value of Malaysian and Indonesian reef ecosystem services. Malaysia has 4,000 square kilometers of coral reefs, valued at RM 145 billion per year. Coral reef related business in Malaysia generates $365 million annually. Indonesia has roughly 85,700 square kilometers of coral reefs, giving it 14 percent of the world's coral reefs (Burke et al. 2002; Dirhamsyah 2005). Value estimates show that extensive damage to Indonesian reefs can cost stakeholders $33,900 per kilometer of damaged reef (Pet-Soede et al. 1999).

4.3 Indonesian Case Sites: Co-managed MPAs

This book examines three cases of co-managed coral reefs in three Balinese villages Lovina, Pemuteran and Amed located in the planned Bali Network MPA in the Southwest corner of the Coral Triangle Region (Figure 4.2).

This MPA is the result of a 2011 scientific assessment of Balinese biodiversity conducted by Conservation International Indonesia and its partners compiled into a report entitled *Bali Marine Rapid Assessment Program 2011* (Mustika 2011). The MPA is also the result of planning, legislation and collaboration between government agencies such as the Marine Affairs and

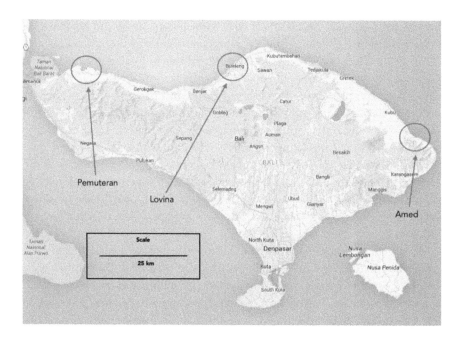

Figure 4.2 Indonesian field sites in the province of Bali.

Fisheries Agency of Bali Province, the Bali Natural Resources Agency (KDSA), several universities and various NGOs. Given Bali's reputation as a global tourism hot spot, balancing the sustainability of its coastal resources with rapid coastal development is a major challenge for government. Thus, government, the private sector and NGOs worked together on a 20-year spatial plan to manage coastal resources known as Local Regulation 16/2009. These MPAs are a component of this long-term plan (Mustika 2011). Within the 563,000 hectare Bali province, there were 25 sites in 9 regions labeled as biodiversity priorities by local government and its NGO partners, with this book examining villages in three regencies constituting priority sites: East Buleleng, Central Buleleng and Karangassem (Mustika 2011).

The first site, Lovina, is in East Buleleng, and is actually the name for an area of several small adjacent villages, including Kaliasem, Kalibukbuk, Anturan and Tukadmungga. Pemuteran is located to the west of Lovina along the same stretch of highway across the northern coast of Bali, and is itself situated in Northwest Bali. Amed in the Karangassem regency in the Northeast of Bali is also a name given to a small area with several villages that include Jemeluk, Bunutan, Lipah, Selang, Banyuning and Aas.

Case study sites in Indonesia all have MPAs visible from the shore, economies that transitioned into reef tourism in the 1990s, bans on commercial fishing on reefs and co-managed MPAs. All three villages have MPAs that impose strict no-take rules on their reefs. Pemuteran, Amed and Lovina were primarily fishing villages before their tourism booms in the late 1980s and early 1990s. They transitioned into reef-based tourism villages in the 1990s. Currently, there is no large- or small-scale commercial or recreational fishing allowed on the nearshore reefs, though random instances of poaching do occur. All locally owned fishing vessels are required to travel several miles off shore when fishing, often bringing them into Javanese waters. Thus, the fishing sector and its management institutions do not have conflicts of interest with the local MPA, though older stakeholders suggested this was not the case in the early years of the MPAs. In these villages, graduated sanctions are in place to prevent fishing and to deter breaches of rules. Line and pole fishing by native villagers is permitted for family consumption only, and was observed in both in Lovina and in Pemuteran.

All three communities have similar timelines of reef destruction and recovery due to similar anthropogenic drivers in their development histories. Destructive fishing methods such as cyanide poisoning and dynamiting were commonly used in Amed, Lovina and Pemuteran from the 1950s up to the 1990s, and it has continued in lesser and varying degrees throughout Indonesia into the 2000s (Erdman 2001). In 1985, *Indonesian Fisheries Law Number 9* prohibited the use of such methods (Badruddin and Gillet 1996;

Pet-Soede et al. 1999). Aside from an uptick in illegal dynamiting in the region during the Asian financial crisis of 1998, illegal destructive fishing has declined substantially. Since the 1980s and 1990s, there has been the greatest decline in users of destructive fishing methods who target their home village reefs. Instead, poachers tend to be from distant locations, including the Philippines and Thailand. Spare a few isolated incidents, the blasting and cyanide fishing have virtually ceased in these case sites thanks to severe legal penalties, increased local awareness, enforcement and graduated sanctions. The case sites experienced similar levels of blast fishing destruction early in their development history since they are nearby one another, approximately 45 kilometers apart. They also implemented no-take rules on local reefs within three years of one another. Thus, this dissertation assumes that the baseline of damages across sites were comparable when management institutions were enacted.

4.3.1 Lovina

Lovina is actually the name for several adjacent villages that make most of their revenue from dolphin watching and reef snorkel tours. Surveys were given in the four villages depicted in Figure 4.3.

Lovina's reefs are visible in near shore waters, with the healthiest reef just off Anturan village. This reef is open to all of Lovina's reef tourism workers to bring boats of tourists for snorkeling trips, regardless of their home village. In addition to reef tourism, Lovina depends on its well-known dolphin-watching tours that bring around 30,000 tourists annually, 60 percent of them from the

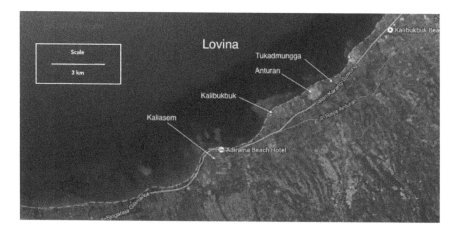

Figure 4.3 Lovina.

Figure 4.4 Traditional jukung boats in Lovina for sailing, tourism and fishing.

West, to observe dwarf spinner dolphins (*Stenella longirostris roseiventris*) (Mustika 2011). Dwarf spinners congregate in Lovina to forage its near shore reef fisheries. Tourists inject almost $10 million annually into Lovina's economy, with nearly half spent on the dolphin watching tours directly. Lovina also depends on snorkel and dive based tourism, with reefs that are still recovering from heavy damage from dynamiters prior to the 1990s.

Stakeholders in Lovina reef tourism include the following: (1) MPA managers, namely the reef and dolphin tourism organizations. These village level reef management institutions manage access to dolphin and reef snorkel tours, and their Balinese names are *Bhakti Dharma Segara* and *Kharya Bhakti Samudra* among others. All stakeholders involved in any form of reef tourism, such as boat captains, guides and engine repairmen are required to join these organizations. (2) Dive industry workers: most dive shops in the Lovina area took divers off-site, but diving and snorkeling does occur on the reefs of Lovina. Included in this stakeholder group are shops that sell snorkel and dive tours. (3) Various tourism workers such as hotel owners, restaurant owners, recreational equipment rental shop owners, and their employees. (4) Village-level governments, such as the head of the village, are involved in

Figure 4.5 A Lovina tourism worker, offering a variety of rentals and activities.

decision making in the reef management institution. (5) International NGOs: There was evidence throughout Lovina that international NGOs had various projects in place having to do with either dolphin or reef conservation. The four main NGOs operating in Indonesia include The Nature Conservancy (TNC), Conservation International (CI), Reef Check and the World Wildlife Foundation (WWF). TNC and its Bali-based Coral Triangle Center assisted the Indonesian government in proposing and surveying reef sites for its MPA network described earlier.

MPA management, rule-making, monitoring, enforcement and conflict resolution is the responsibility of the reef and dolphin tourism organizations. Village-level government assists with financial arrangements and sanctions delivery if necessary. NGOs also assist with ecological monitoring.

4.3.2 Pemuteran

Pemuteran is a village with a relatively large area of coral reefs. It generates most of its revenue through dive tourism, with its most popular reefs identified in Figure 4.6.

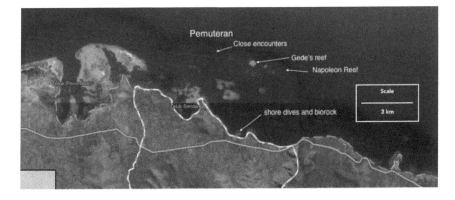

Figure 4.6 Pemuteran. *Source*: Google Maps.

Figure 4.7 The Pemuteran MPA.

Pemuteran was one of Bali's poorest villages up until 1990. Since 1989, its land prices have soared to 400 times what they once were (Savitri 2001). Respondents suggested that this figure is many times higher in 2013 though data is not available. Currently, its economy is expanding as the village gains reputation for being a world-class, award-winning destination for reef

Figure 4.8 A tourism worker's business in Pemuteran.

tourism. Besides its natural reefs, it has several Biorock formations with mineral accretion technology, or charged wires that increase coral growth. These formations are free for tourists to visit, but NGOs that maintain them ask for a donation of $2 added onto the dive price voluntarily. Biorock sites are within a short swim from the shoreline. Facilities for environmental education, such as libraries and information centers line the shore. Tourists are encouraged to learn about how the community acts as a reef steward.

Stakeholders in Pemuteran include (1) The MPA managers, called the *Pecelan Laut*: the religiously inspired sea police who patrol the reef on Saturdays and Sundays and are the primary enforcers of the community-based MPA. Hierarchically organized, members have different management responsibilities at different levels and membership is highly prestigious in the community. (2) The local hotel owners: these owners of medium to large hotels have made investments in reef management organizations, spearheaded their own conservation projects, sponsored and organized cleanups and coral restoration. They are well known in the community. (3) Village-level government: this includes traditional village leadership (i.e., the head of the village plus an elected and appointed staff). (4) Local NGOs: NGOs are primarily responsible

Figure 4.9 An MPA manager and NGO worker in Pemuteran with wire that will be hooked up to a device under water that will be charged, and ultimately will be used to grow Biorock. One of these nameplates costs a tourist visitor around $40.

for invasive species removal, education and biological monitoring. (5) Global NGOs: these are widely known NGOs such as Conservation International. Global NGOs primarily assist the community with ecological monitoring, and work to scale up MPA management across Bali. (6) Dive industry workers: these stakeholders can range from owners of dive shops to diving employees to maintenance crews and boat captains. (7) Various tourism workers: this catch-all group includes anyone working in tourism but not necessarily in the dive industry, including boat captains, restaurant owners, waters, hotel staff and tour package sale point operators.

MPA management, rule-making, monitoring, enforcement and conflict resolution is the responsibility of the Pecelan Laut. Like Lovina, the village-level government assists with financial arrangements and sanctions delivery if necessary. Local NGOs maintain Biorock, remove invasive species and build artificial reef formations. Global NGOs support local NGOs with financing and assist in scaling up co-management efforts across Bali. They also provide ecological monitoring support.

Figure 4.10 Signage visible throughout Pemuteran on the community-based MPA.

4.3.3 Amed

Amed, located in East Bali, is globally recognized by diving enthusiasts for its famous dive sites, including, most notably, the USS Liberty Wreck, with many of its other well-known sites depicted in Figure 4.11.

Its economy is based almost entirely on dive tourism. Prior to the 1990s, most of Amed's residents were fishermen, who then underwent livelihood transitions from fishing to tourism in the 1990s and early 2000s. In 2016, villagers engaged primarily in reef tourism can still be known to fish in a part-time capacity, and traditional *jukung* boats line the shores, often parked in front of small bungalows owned as a primary sources of income for many households. Fishing is banned within two kilometers of the shore, and only line and pole fishing by local stakeholders is allowed on reefs.

There is strongly enforced payment system implemented by the village reef management institutions. Every boat that uses a dive site is required to pay the village that manages that particular dive site or reef. For example, if a bus brings a group of divers on a day trip to Jemeluk Bay, the dive operator owes the village reef management institution a small payment for each diver.

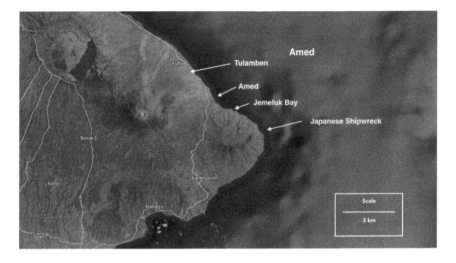

Figure 4.11 Amed.

Additionally, dive shops, even if they are based in Amed, owe the villagers who administer reefs and dive sites parking fees, reef access fees and fees for women who carry all of the diving equipment from cars and shops to the shoreline for the dive operators. It is not uncommon to see women carrying four tanks at one time. These women tend to not hold any formal education, be middle aged and older, married to men who once worked as fishermen and only speak Balinese. A very small portion of Amed's residents, primarily older men, still engage in fishing for all of their livelihoods, depending on tuna and barracuda that they sell to local restaurants. Many fishermen speak of declining stocks and having to go further and further to find fish, making tourism an increasingly attractive option.

Stakeholders in Amed include the following: (1) MPA managers: these are members of organizations usually formed by fishermen's cooperatives. They have monthly meetings, rules for reef access and tourism, rules for local dive shops to follow, payment collection agents that collect fees at dive sites and conflict resolution mechanisms. (2) Village-level government: this includes the headman and secretarial positions. Village government has a wide range of responsibilities including approving development in Amed and financing certain conservation programs. (3) Dive industry workers: The dive industry has grown exponentially in Amed since the 1990s. There is an even mix of expatriate owned dive shops and Indonesian-owned dive shops, with rules that dictate an Indonesian must legally own 51 percent of any dive operation. This often means that expatriate owners will find a trusted Indonesian friend who owns the business in name only. Law reserves certain positions in the

Figure 4.12 The Amed MPA.

dive industry for Indonesians only, such as the divemaster title. This law is often broken. Women porters are a pivotal component of the dive industry with livelihoods guaranteed by village level government regulations. This is a very lucrative position, earning $1 per tank, women can earn 10–20 dollars a day for porter work, compared to fishing when they could earn a fraction of that. (4) Various tourism workers: many households in Amed own small bungalows, many operate on small budgets and do not offer amenities such as air conditioning and hot water. Many households in Amed work in the tourism sector in restaurants, transportation, hotels or dive shops. Boatmen, for instance, are stakeholders in the midpoint between transitioning into the tourism sector and out of the fishing sector. They tend to earn most of their living from taking divers to dive sites, or doing sunset cruises, but will fish part time if they need to. They tend to speak less English and are typically older than other locals who work in the dive industry. (5) International NGOs: There was evidence throughout Amed, such as the signs at every dive site, that international NGOs had projects in place on reef conservation or ecological monitoring for disturbances such as coral bleaching.

MPA management, rule-making, monitoring, enforcement and conflict resolution is the responsibility of the village-based MPA management institution.

Figure 4.13 Women porters in Amed's dive industry.

Village-level government assists with financial arrangements, collections of fees at dive sites paid by users and sanctions delivery, if necessary. Global NGOs also assist with ecological monitoring.

4.4 Malaysian Case Sites: Centrally Managed MPAs

This book examines two cases of centrally managed MPAs in villages on islands that fall within MPAs, Kampung Palau Perhentian on the Perhentian Islands and Tekek village on Tioman Island (see Figure 4.16). Both case sites fall within the government of Malaysia's MPA networks that it began to form in the 1980s in order to protect its coral reefs. These MPAs are both "no-take," meaning that fishing is prohibited within two nautical miles of the coastline. Both the Tioman Island and Perhentian Island MPAs were established in 1994 under the *Fisheries Act 1985* for Malaysia to fulfill its obligations under the CBD and its national implementation plan create MPAs. A central focus of Malaysia's CBD implementation plan is conserving its biodiversity while still allowing economic opportunity for nearby villages. Both villages were originally small fishing villages that have since transitioned into

Figure 4.14 Sign indicating the "Fishermen's Cooperative PERNETUS village meeting point." The men who form the fishermen's cooperative are also tasked with roles as MPA managers.

larger economies dominated by reef and island tourism. Each year, increasing numbers of visitors come to these MPAs for activities ranging from diving to jungle trekking. Much like the Indonesian field sites, the mid-1990s saw tourism expand rapidly as well as infrastructure projects within both MPAs.

Similar to the Indonesian sites examined in this study, the Malaysian communities studied in this book have reefs visible from the shore, economies that transitioned to tourism-based in the 1990s and bans on commercial fishing; yet, unlike Indonesia, reefs are centrally managed. Many locals in the villages still do participate in fishing, but do so outside of the MPA boundaries, and use it as a supplementary income after tourism. There are graduated sanctions in place to deter rule breaking among locals.

Patterns of degradation and policy interventions are somewhat similar to the Indonesian sites. Prior to the implementation of MPAs, destructive fishing methods such as bombing, as well as fishing practices that cause excess damage to reefs such as the use of trawls and nets were common on the reefs of Malaysia. Implementing the MPAs ended the most brazen acts of destructive fishing for the most part due to the expansion of community awareness on

Figure 4.15 Dive industry workers from Amed wearing traditional dress for Independence Day ceremony commemorating the launch of a conservation program where dive industry stakeholders obtain training on ecological monitoring.

the value of reefs and the potential for reef tourism. Like the Indonesian sites, contemporary poaching violations tend to come from foreign vessels from Thailand or China.

New conservation problems emerged after the formation of the MPAs such as coral damage caused by divers and snorkelers beginning in the early 2000s. Additionally, MPA authorities have always dealt with increased poaching during the northeast monsoon season from October to February, and continue to do so. This led to the Malaysian government to ask the United Nations Development Programme (UNDP) to assist in building the capacity of its marine parks to stop the new forms of damage caused by tourists, as well as to aid in the livelihood transitions of villagers from the fishing sector to tourism (UNDP 2013). A key part of this program involved helping marine parks officers increase patrols searching for vessels violating the fishing bans.

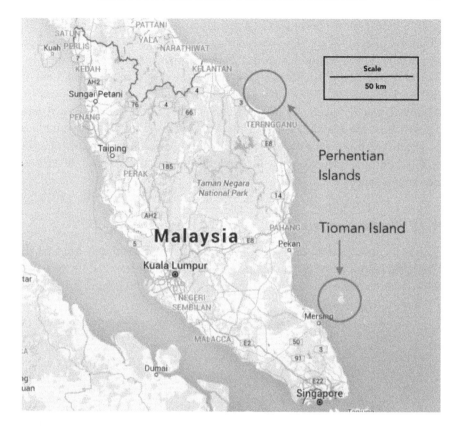

Figure 4.16 Malaysian MPA field sites.

Additionally, villagers were educated on reef ecology and proper behavior on reefs plus training on recognizing rule breaking and how to report it to authorities. By 2008, this led to increased numbers of arrests for rule breakers, poachers and violators caught fishing or damaging coral (UNDP 2013). Since 2011, according to UNDP reports, the project's success has become evident since violations have decreased (UNDP 2013).

Malaysian MPAs have sought multilateral support in addressing several challenges including the need for bureaucratic streamlining and issues with under staffing. In 2004, the Malaysian government transferred the management of its MPAs from the Department of Fisheries Malaysia to the Department of Marine Parks Malaysia in order to (1) promote sustainable resource management and (2) increase tourism to the MPAs. Urban planning and land use decision-making powers on the islands themselves belong to the state governments. Poor land-use planning and waste management has been a considerable driver in the degradation of the reefs (Harborne et al. 2000; Islam et al. 2013). The bureaucratic reshuffling was meant to increase interagency collaboration between state governments and the Department of Marine Parks. Malaysian Marine Parks have always been understaffed (UNDP 2013). There are currently approximately 100 staff members employed by Marine Parks, with roughly a dozen working in the central office in Kuala Lumpur and the remaining staff in the MPAs themselves. Capacity-building programs, such as that by the UNDP, have sought to train Marine Parks officers to focus on monitoring to allow for a relatively small staff to have the largest potential impact.

4.4.1 Perhentian Islands

The Perhentian Islands, commonly referred to as the Perhentians, are a small cluster of islands in the South China Sea nearly 19 kilometers off the coast of Malaysia, reachable by boat from the state of Terengganu (Figure 4.17).

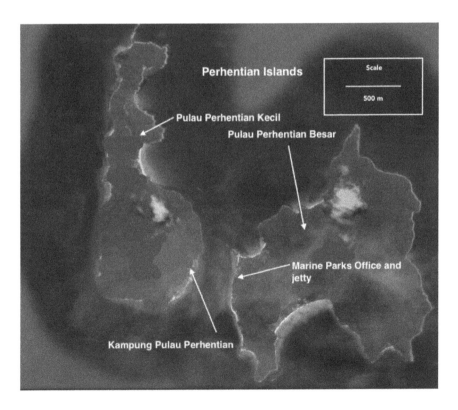

Figure 4.17 Perhentian Islands MPA.

Figure 4.18 Water taxi operators on Pulau Perhentian Besar.

Two of the Perhentian Islands have permanent settlements and infrastructure for residents and tourists, including Pulau Perhentian Besar (Big Perhentian) and Pulau Perhentian Kecil (Small Perhentian). Several reefs popular among tourists are visible from the village, while many other popular sites require a boat ride often less than 10 minutes in duration. Kampung Pulau Perhentian, the village, is located on Pulau Perhentian Kecil and has a population of 2,000–2,500 inhabitants, two large mosques, three jetties, a school and a small clinic.

The majority of inhabitants make their living in the tourism sector, with diving as the dominant reason for attracting tourists. Prior to the 1990s, most of the inhabitants of the village were fishermen, with economic transition occurring at a rapid pace. Many working in dive tourism in the Perhentians come from mainland Malaysia or foreign countries. Very few are native to Kampung Palau Perhentian. There are approximately 50 hotels on the islands, ranging from backpacker lodges and dormitories to high-end resorts. Although some owners are from the village, most are from mainland Malaysia or foreign countries. The same is true of dive centers. Villagers do tend to own restaurants, convenience stores and water sports rental shops on the islands. There are no paved roads and all travel occurs via water taxis (Figure 4.18).

Figure 4.19 Perhentian Islands MPA.

Stakeholders in Kampung Pulau Perhentian include the following: (1) Marine Parks officers: Men and women are tasked with management, monitoring, enforcement and educational duties within the Marine Parks system. They work from large, prominent, centrally located offices in most Marine Parks, typically next to the main jetty. (2) Dive industry workers: dive industry workers can range from dive shop owners to diving employees such as instructors to tank-fillers and boat operators. (3) International NGOs: NGOs such as Reef Check have an important presence in MPAs, and are often spoken of as equally important to Marine Parks officers in conservation. They typically implement ecological monitoring and public outreach. (4) Various tourism workers: this catch-all term includes those working in hotels, restaurants or recreational businesses such as water sports rental locations (Figure 4.20). (5) State governments: state government offices hold planning and land-use decision-making power on the islands within MPAs. There was no office or personnel located on the island itself. (6) Local universities: local universities, much like NGOs, are authorized by Marine Parks through permit schemes to conduct biological monitoring surveys in the MPA. During this study, I observed students from the University of Malaysia Terengganu performing

Figure 4.20 A dive industry worker from Kampung Pulau Perhentian Besar.

ecological surveys. MPA management, rule-making, monitoring, enforcement and conflict resolution is the responsibility of the Malaysian Marine Parks. NGOs and universities help with compiling scientific reports and data, but they must apply for a permit in order to do so.

4.4.2 Tioman Island

Tioman Island is the largest, and the only inhabited, island of the 14 islands included in the Tioman Island MPA. The Tioman island MPA is located 59 kilometers off the coast of the state of Pahang. The MPA was established in 1994 under the Fisheries Act 1985, similar to the Perhentians. Tioman is home to nine small villages, the largest being Tekek, and has 3,000–4,000 inhabitants (Figure 4.23). There are many reefs popular among tourists that can be seen

Figure 4.21 View of Kampung Pulau Perhentian.

Figure 4.22 View of Kampung Pulau Perhentian.

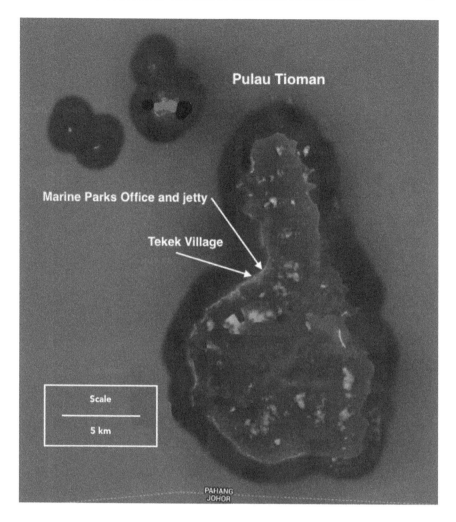

Figure 4.23 Tioman Island.

from the shore, but there are also dozens of popular sites that require a 10–20 minute boat ride.

Most of the tourism infrastructure is located by the jetty and on the Western side of the Island from Air Batang to Tekek Village. There is one small road on the West side of the island, but travel elsewhere requires a water taxi. Of the Malaysian Marine Parks, Tioman is the most commercially developed, largely because state officials have sought to expand its jetty for yacht tourism. Tioman has approximately 40 hotels ranging from backpacker to luxury, much like the Perhentians. Tioman villagers own very few hotels and dive shops

Figure 4.24 Dive industry workers, Tioman Island.

(Figure 4.24). Instead, owners tend to come from mainland Malaysia or foreign countries. Locals do, however, own and operate many of the island's smaller restaurants, convenience shops and watersports rental stands.

Tioman Island's economy prior to the 1990s was historically dominated by fishing and is now almost entirely based on tourism. Fishing was banned with the creation of the MPA and residents were left with no choice but to transition out of the sector. Many former fishermen now make their living acting as water taxi drivers, snorkel tour operators or boatmen who operate the speedboats between the mainland and Tioman. Many working in dive tourism in Tioman come from mainland Malaysia or foreign countries. Very few are native to Tioman.

Stakeholders on Tioman Island are identical to those of the Perhentian Islands described above. There is one addition, however: local NGOs. These NGOs work on issues that fall into gray areas of governance over which state and federal agencies have been unable to address. These include waste removal and dealing with sewage on the islands. There is a large amount of cooperation between those working in international NGOs and those working in local NGOs.

Figure 4.25 Display inside Marine Parks office, Tioman Island.

4.5 Controlling for Differences across Case Sites

One of the most important considerations in writing this book comparing Indonesian and Malaysian MPAs was setting up the comparison by controlling for key differences across field sites. I selected case sites controlling for cultural differences, the presence of civil society, economic development trajectories and geography. Although the island of Bali, where all of the Indonesian sites in this study are located, is predominantly Hindu and Malaysia is a Muslim country, there are undeniable cultural similarities between these Southeast Asian countries. For example, Indonesia and Malaysia share a language and many cultural attributes due to their proximity and ever changing borders through their closely linked histories (Kahn 1998). Additionally, both Indonesia and Malaysia have the presence of an active civil society.

Economic development patterns across sites have key similarities. In all case study sites, households depend primarily on the reef for marine tourism industry. Both Indonesia and Malaysia are experiencing economic growth at similar rates, with annual growth in Malaysia at 6 percent and in Indonesia at 5 percent. Per capita income at both sites was comparable and primarily based on income from tourism, whereby Bali, as of 2009, has a per capita income of 12 million Indonesian Rupiah (IDR), or $1200, and the states of Terrengganu (where the Perhentian Islands are located) and in Pahang (where Tioman Island is located) have per capita incomes of 23,285 Malaysian Ringgit (MYR) and 26,759 MYR, or $5904 and $6785, respectively (Erviani 2009; Department of Marine Parks Malaysia 2014; Bali Government Tourism

Office 2015). Other places in Indonesia with tourism economies based on reefs have far lower per capita income when compared to Bali, such as sites in Sulawesi and Papua, causing those cases to be eliminated (Ananta et al. 2011).

The World Bank classifies Indonesia as a lower middle-income country, and Malaysia as an upper middle-income country (World Bank 2015a,b). However, given that 1.9 million people visit Bali each year, generating $2–3 billion per year in its tourism economy, Bali and Malaysia make for a closer comparison when controlling for tourism economies compared to other islands in Indonesia. Poverty levels in Indonesia and Malaysia are quite different, 11.3 percent and 1.7 percent, respectively, but poverty levels in Bali, at 6 percent, are roughly half that of the national estimates for Indonesia, making them close enough for a reasonable comparison (Erviani 2009).

One factor that this study did not initially control for, because data does not exist, is the amount of visitors that engage in reef tourism at each site. Through interviews with high-level experts and the use of official visitors statistics that show only hotel visits, I determined that Malaysian sites in the study see roughly 27 percent fewer tourists compared to the Balinese sites. Interviews with government officials and those working in tourism and regional development suggest that the difference is due to the more conservative elements of Malaysian society and the social restrictions on dress and behavior, such as the difficulty in freely obtaining alcohol, that limit the numbers of tourists compared to the less conservative Bali.

Tourism is thought to be increasing in both Balinese and Malaysian sites, with data on trends available in the Malaysian case and estimates from interviews with officials in the Balinese case. The number of visitors to Tioman, according to the Department of Marine Parks Malaysia (2014), was 263,701 in 2014. This is a 31.5 percent increase in tourist numbers over a 10-year period. The number of visitors to the Perhentians and nearby Redang Island was 262,094 in 2014. The Perhentians saw a 79 percent increase in visitors in a 10-year period.

Indonesia and Malaysia are ranked 17th and 23rd, respectively, in terms of the contributions that tourism makes directly to their GDPs (WTTC 2015). The Organization for Economic Cooperation and Development (OECD) reports on global tourism statistics and includes data for Indonesia overall. The OECD finds that on average, tourists who visit Indonesia spend $1948 and spend a week in country (OECD 2014). Buleleng regency in Bali where Lovina and Pemuteran are situated saw 666,776 visitors in 2014. Karangasem, where Amed is located, saw 423,740 visitors in 2014 (Bali Government Tourism Office 2015). The Bali visitor numbers are more than likely a low estimate because of the high popularity of day trips to Buleleng and Karangasem

Table 4.1 Economic and demographic overview of Indonesia and Malaysia. Percentage of GDP from tourism from ETN Global Tourism News (2013) and Christie et al. (2003) for Indonesia and Malysia, respectively.

	Indonesia	Malaysia
Population	249.9 million	29.72 million
GDP per capita	$3475	$10,538
GDP	$301 billion	$867 billion
% of GDP from tourism	12.5	11
Annual GDP growth	5.6%	6.2%
Total visitors/year	8 million	27 million
Government	Republic	Constitutional monarchy
Independence	1949	1957
GINI coefficient	46.2	36.8
Literacy rate	93.1%	92.9%
Land area	1,904,000 km^2	330,000 km^2
Coastline	55,000 km	5,000 km
EEZ	6,159,000 km^2	335,000 km^2

from the main tourist centers of Denpasar and Badung. Limited and informal polls at several of the most popular dive sites including Tulamben and the Japanese Wreck site near Amed puts 10–40 percent of divers and snorkelers as "day trip" visitors who are not counted in these numbers. From these visitation statistics, a rough estimate of the 27.6 percent difference between Malaysian and Indonesian sites was derived and then verified with experts during fieldwork.

Table 4.1 presents other factors considered important when selecting Malaysia and Indonesia as countries where comparison is possible. Similar levels of inequality, large coastlines and exclusive economic zones and similar economic dependencies on tourism make Indonesia and Malaysia valid sites to compare.

4.6 Ecological Results: Overview of Coral Cover Results

The results of the living coral cover surveys of the five field sites across Indonesia and Malaysia are reviewed in this section. Overall, this study found no significant differences in living coral cover after statistical analyses between field sites in Indonesia and Malaysia. Concluding that centrally managed and decentralized institutions have similar impacts on ecological outputs may not be possible however, since it was also discovered that Indonesia receives 27 percent more visitors to its MPAs compared to Malaysia. This study also found through dozens of observations that divers and dive tourism does cause damage and stress to corals, through touching, trampling and sedimentation. Thus, given that the Indonesian reef system is not statistically different from the Malaysian reef system despite the greater volume of visitors and stress to the reef, co-managed institutions could be be one reason why this is the case, though it is unclear how much credit they are due.

For a technical, in-depth look at the findings on coral cover broken down by site, refer to appendix C.

4.7 Summary of Living Coral Cover Findings

Remarkable, above all else, was the amount of diver contact to the reef witnessed countless times (images included in appendix C). Research has been done on dive industry impacts to coral reefs. Some studies have shown that while reefs with highly frequented dive sites tend to have more broken coral, damage is not significantly related to diving intensity (Hawkins et al. 2005). Other studies suggest that diving intensity is indeed significantly related to damage to reefs with most damage coming from divers kicking up sediment and from contact to the reefs which increases when divers are using a camera (Tratalos and Austin 2001; Zakai and Chadwick-Furman 2002). The findings of the latter studies coincide with the author's findings in this research.

Another point worth noting is that while coral cover across the two sites is not significantly different, Bali has seen unprecedented intensity in coastal development including intensive land clearing for a variety of tourism and infrastructure purposes. While the similar things can be said for the Malaysian sites, the scale and intensity in Bali is much greater. For that reason as well we would expect to see reefs exhibiting more signs of disturbance when compared to the reefs of Malaysia. Additionally, Balinese reefs have been experiencing bleaching events since the 1997–1998 El Niño, whereas Malaysian reefs began experiencing large scale bleaching in 2010. Why then are Balinese and Peninsular Malaysian reefs exhibiting similar qualitative levels of coral cover? In the remaining chapters, I argue that social arrangements where local actors have more power to manage reefs and enforce rules offer one compelling explanation for the difference. It is important to note, however, that it is not the only explanation, but instead one that deserves to be studied. For example, ocean upwelling in the Northern parts of Bali can explain reef recovery and resilience from thermal events, and the differences in coral cover may be explained by chance events or biophysical factors.

I argue however that the idea that social arrangements have little or no impact on the ecological outcomes exhibited by reefs too simplistic, but likewise, saying they are the only reason reefs look a certain way after a number of years is equally simplistic. Instead, the case studies offered in this book should be viewed as a small component of a wider story, whereby social arrangements impact ecological outcomes along with a variety of other factors, and it's important to be able to say which arrangements are more effective, which the following chapters seek to do.

Chapter 5

INTEGRATED MANAGEMENT OF MARINE PROTECTED AREAS

5.1 Overview

The following chapters examine the social differences between Indonesian and Malaysian reef management contexts. These differences fall into three categories: integrated management of the social and ecological system, legitimacy and adaptive capacity. In the next three chapters, I will build a foundation for my conclusion that Indonesian reef management institutions, because they do integrated management better, have more legitimacy, and afford their stakeholders greater capacity for adaptive management are better suited to managing reefs in the Southeast Asian context.

Although there were no findings of statistically different levels of coral cover in this research between Indonesia and Malaysia, stakeholder perceptions varied significantly between top-down-managed marine parks in Malaysia and co-managed sites in Bali, Indonesia.

This chapter discusses the different perceptions among marine protected areas (MPAs) stakeholders in Indonesia and Malaysia on integrated management. Integrated management occurs when management institutions create rules for access and extraction of an ecosystem that consider economic activity (such as fishing or building new hotels on the coast) alongside ecological quality indicators that indicate the health of the reef. Integrated management has three parts: *balancing* between stakeholder groups; *limitations* where some stakeholders "win" and some "lose," but with strategies to help those who lose transition into new livelihoods; and what I call *"steam releases"* where some rule breaking can occur by a very small amount of people or in a very limited scale.

To better illustrate this concept, I offer two hypothetical cases of reef management institutions, Case A and Case B. In Case A, there is integrated management. Leaders in the MPA make rules that limit the most ecologically damaging activities to the reef with a careful balancing act in mind. This balancing seeks to prevent all out unemployment of fishermen, while also preserving the health of the reef. For instance, they ban destructive fishing and the use of highly damaging methods such as trawling, while still allowing fishing with nets as well as line and pole 12 kilometers offshore from the reef. Fishing was not banned, but in order to protect the reef and ensure that functional groups that are critical to reef health are not overfished, fishing was moved further offshore and certain gear types were banned. This means that the dive industry stakeholders are winners, and those who primarily would fish the reef for their income are losers. Managers in the MPA in Case A would then need to help the reef fishermen adjust to their lives as fishermen who work further offshore, perhaps by subsidizing new gear, engine and boat repairs that make sure these longer trips are safer, or conducting English classes where the fishermen could act as part-time boat operators for the dive sector. Additionally, since the dive industry stakeholders are the primary beneficiaries, they too could contribute financially to the transition of fishermen from reef fishermen to offshore fishermen. Otherwise, they could offer alternative sources of employment such as vessel operators to fishermen who chose to leave their sector after the reforms.

Because the MPA institution in Case A knows that people in the village like to eat coral trout and drink the local rice wine during parties with friends, they allow people born in the village to take a small amount of fish from the reef using either line or poll or a spear gun so long as (1) they are not seen doing this by tourists in the day and (2) they only take enough for their social activity. This amount can follow a very simple rule of thumb, such as you may only take as much as you can carry with your own hands. This way, the most destructive fishing is banned while fishing in general is not banned, and social forms of reef fishing that make a small impact are permitted in a way that is easy to enforce because a small amount of people do it from time to time, and it happens locally.

The MPA in Case B on the other hand does not manage in an integrated way. Case B has an MPA with a strong and well-liked leader who also did some college work in marine sciences. He sees nearby villages getting small grants and support programs from well-known international nongovernmental organizations (NGOs) and thinks his village can get this type of recognition as well. The men who make up the MPA decision-making structure know that fishing is harming the reef, and they ban all of it. This was easy for them to do, since most people in this village no longer fish, and instead work in tourism. Fishing several kilometers offshore is allowed, but they do not assist former reef fishermen in transitioning into the livelihoods of offshore fishermen or make any attempt to bring them into the dive sector. They do not make the dive sector pay for their preferred status to help out the fishermen who no longer have jobs, and resentment grows between fishermen who had been there earlier and divers who are seen as newcomers. Fishermen begin to notice that many divers damage the reefs and see the fishing ban on reefs as an unfair and arbitrary rule. By cover of darkness, they begin to use trawls and large nets around the reefs in the no-take area. These gear types, when poorly maintained, often come undone and are seen by divers the following morning. People in the village miss the old days when they could send their oldest boy with a pole and line to get a few snappers for dinner on a Saturday night. They see more foreigners in town with dive gear and resentment continues to grow. Fewer and fewer fish are seen each year on reefs, and the dive industry begins to slow as word of mouth travels that other villages have reefs with better protections and more fish. In the MPA in Case B, there was no balancing, no acknowledgement of winners and losers, no push to help losers transition into new livelihoods, and no steam releases that preserved social fishing patterns.

For this section on integrated management, I asked reef stakeholders questions focused on linking livelihoods and the conservation efforts of MPA managers in order to determine whether conservation and local economic activities were seen as compatible and mutually beneficial. Whereas stakeholders in co-managed settings believed that conservation and livelihoods were closely linked, those in top down MPAs did not. Half of respondents in Malaysian MPAs and 73 percent of Indonesian stakeholders felt that businesses on the islands invest in conservation. Nearly 80 percent of Malaysian stakeholders said that Marine Parks does not help them earn a living off reefs, while nearly 80 percent of Indonesian stakeholders said their MPA did help them earn a living off reefs. Stakeholders in both sites felt that NGOs were a major source of help in the community for linking the tourism business, ecological management and conservation. Based on these stakeholder perceptions, I find that co-managed systems are more successful at integrated management.

5.2 Survey Results

The results in Figure 5.1 show the statistical tests that I completed on three questions meant to query stakeholders on integrated management of social, ecological and economic systems. Every question except for the last one had significantly different responses between Indonesian and Malaysian sites.

5.3 Different Perceptions on Conservation and Livelihood Links

As mentioned above, integrated management is a balancing act where limits on ecological impacts are introduced that affect some people in some livelihoods more than others. Those who benefit along with MPA managers can act to

Do you see conservation and your livelihood as being linked?

	No	Yes	Total
Malaysia	84 (0.83)	17 (0.17)	101
Indonesia	22 (0.14)	138 (0.83)	160

$$p = 0.000***, z = -11.12$$

Do reef tourism businesses invest in projects that help conservation?

	No	Yes	Total
Malaysia	53 (0.51)	51 (0.49)	104
Indonesia	42 (0.26)	118 (0.73)	160

$$p = 0.00***, z = -4.08$$

Does the reef management organization respond to crises (such as illegal fishing) in a way that helps you continue to earn a living based on reef tourism?

	No	Yes	Total
Malaysia	74 (0.88)	10 (0.12)	84
Indonesia	31 (0.19)	129 (0.81)	160

$$p = 0.00***, z = -10.30$$

Do NGOs, either local ones or international ones, support you in conservation?

	No	Yes	Total
Malaysia	24 (0.22)	79 (0.77)	103
Indonesia	41 (0.27)	101 (0.66)	152

$$p = 0.50, z = 0.66$$

Figure 5.1 Survey responses to questions on integrated management. *** : $p < 0.001$.

support those who do not. In Case A, the MPA and the dive industry offered livelihood alternatives to those forced to abandon reef fishing. In Case B, there was no acknowledgement that fishermen had to sacrifice their livelihood while the dive industry benefitted. This created resentment and rule breaking. In order for any form of integrated management to occur, there must be some perception among stakeholders that conservation and their livelihoods are linked, or else any rules that cost individuals money, limit fishing or limit hotel construction will not be followed. This study found significant differences in stakeholder perceptions on the links between livelihoods and conservation.

5.3.1 Malaysia: Conservation is not my problem

Stakeholders across economic sectors in Indonesian co-managed reefs (those working in hospitality, as shop owners, in transportation or in the dive industry) saw conservation as their responsibility one way or another. This widely held belief was not present in Malaysian Marine Parks sites except the dive industry. Nearly universally, dive industry stakeholders felt that conservation was indeed their responsibility, but that any attempts to do so could be actively impeded by Marine Parks officers, or even bring them unwanted attention from the authorities.

Many other stakeholders in Malaysian Marine Parks, even those who make their living directly on the reef—boatmen, for example—did not see the connection between conservation and their livelihoods. "I do not think that it is my job to take trash off the reef," said Ali, a boatman leading a snorkel tour on the Perhentian Islands. "That is why we have Marine Parks officers." He then threw his lunch over the side of the boat, plastic wrapper and all. A younger boatmen threw his cigarette in the water over a very popular snorkeling point and said, "It is not our job to check the reefs and clean them, this is the job of Marine Parks. It is not my problem."

This was not an uncommon opinion in Malaysian Marine Parks, where many outside of the dive industry felt that locals need not intervene at all for conservation because it is the responsibility of government. When interviewing those who worked in hospitality, shops or transportation, the opinions were even more pronounced that the reef has nothing to do with them and that conservation is strictly the job of government. The further away stakeholders got from day-to-day work on the reefs, the less they felt that conservation impacted them. In the words of a long-time restaurant owner on the Perhentian Islands, "I do not worry about conservation. Marine Parks people are paid to do that. It doesn't affect me." When I asked him whether he thought fewer people might come in the future if the ecosystem was degraded, he replied, "We don't think like that here. I think day-to-day. I count my register at night and hope for a good day tomorrow. I have many children and they have school fees."

The main reason that Malaysian stakeholders outside of the dive industry did not link conservation and livelihoods was a lack of education and a low income. Repeatedly, respondents mentioned not having enough time, money or resources to properly dispose of rubbish or engage in other individual behaviors either at home or at their businesses that would protect the reef. Many mentioned that they were more worried about earning an income to feed their families than they were about the long-term health of the reef. An NGO in the Perhentian Islands implemented a recycling program. When villagers were asked about whether they thought this program helped their incomes, they responded that it might, but they were not sure how. "Maybe Western tourists like to recycle so it is good for business here," one respondent said. To stakeholders, the link between a healthy ecosystem, visitors and their own businesses was weak.

Several respondents suggested that Marine Parks needed to do a better job of teaching residents of the villages that conservation and livelihoods were linked. "All that they do over there in Marine Parks is make posters and hang them in hotels," said one dive industry worker echoing the sentiments of many respondents. "Why don't they teach the local residents that when they throw trash into the ocean, people won't come to their roti canai shop next year?" Several stakeholders had similar criticisms about Marine Parks and how, in their view, the only real management work that Marine Parks officers did included hanging informational posters and distributing printed materials. Many questioned why it was that this information was not more widely distributed. "It is not like they are every patrolling for illegal fishermen," said one respondent. "So why can't they go out and educate the villagers on how their behavior impacts the environment?" Another stakeholder had the following to say, "I see boatmen who bring people out to snorkel tours drop anchors onto the reef at least once a month. If I've seen it, Marine Parks has seen it too. Why aren't they training the boatmen?" he asked. He continued, "I don't think it would be very hard or very expensive to train the water taxi guys and small boat snorkel tour guys not to throw anchors on the reefs, I don't understand why it doesn't happen."

Several respondents argued that Marine Parks officers are largely absent from the villages where they work and from their places of employment in general. "Marine Parks officers are never around," said a long-time dive shop employee on Tioman Island. "So how would they do community engagement work?" Another added, "I would say the community around here has no idea who these guys are, and if they tried to run educational programs telling people, 'don't touch this' and 'don't do that' that people wouldn't listen to them." Another respondent added, "In order to educate the community, you need to actually be in the community. People need to know who you are and trust you. That is not Marine Parks." The idea that Marine Parks officers were frequently absent was the most pronounced reason why stakeholders felt that they lacked the capacity to engage community members and train them in the link between the environment and the ecosystem. One stakeholder added, "I wonder if

Marine Parks officers think that reef health is related to their own job security. Given their work ethic, and the fact that I never see them, I think they do not."

Respondents who worked in the Marine Parks contested these views on their lack of community engagement. "We have programs every year in Kampung Pulau Perhentian for Independence Day. We have a fishing contest and we train them that reef health is important for their village economy." Many other respondents questioned the choice of a fishing contest when linking ecosystem health and livelihoods. "Why a fishing contest and not a rubbish cleanup?" asked one NGO worker. "This is really typical of Marine Parks. Often they don't make any sense." Other Marine Parks employees noted that the United Nations Development Programme (UNDP) had conducted a long training program on community engagement and that they had noticed a large local mindset change in recent years. "People understand that healthy reefs mean healthy village economies," said one Marine Parks employee. "You would not have seen this level of education ten years ago. It is a long process, and it isn't perfect, but the mindset is changing." Some Marine Parks officers said that the mere presence of the MPA and the popularity as a tourist destination by Malaysians and foreigners was itself a major source of local mindset change. "People come here and spend a lot of money on holiday. Villagers see this, and they think, 'these people are coming for reefs' and they work together to save the reefs. Maybe it isn't perfect but is it happening."

5.3.2 Indonesia: The reef economy

In co-managed MPAs, 83 percent of stakeholders felt that their livelihoods were linked to reef health and conservation. This opinion spanned age, economic sector and job description. Hotel owners, restaurant owners, cab drivers and housewives all felt that a healthy reef made them wealthier. This perception was very widely held in all villages studied for this book. Indonesian stakeholders linked the reef and their income commonly citing two themes: the links between reef health and wealth and the importance of leaving reefs as an opportunity to future generations.

Nearly every respondent spoke of reef systems as synonymous with livelihoods. "Without reefs, I could never have earned enough to buy my hotel," said Ketut, a hotel owner in Pemuteran who has a newborn, two children and a small hotel with three rooms. "My father was a fisherman. He would make [1 dollar] a day selling fish to market, and my mother sold salt. I made [20–30 dollars] a week giving sunset tours on my father's fishing boat, I saved, and built this hotel." Stories like this were common in co-managed sites. Ketut continued, "Without these reefs people like you would not come. Without these reefs, I would earn a dollar a day like my father did."

A boatman in Amed, Gede, told me a similar story. "My family were farmers and fishermen before the tourists came," he said. "Maybe 20 years ago people noticed our reefs and they began to come. These reefs are the reason I can fill my boat with tourists for sunset and earn [20 dollars] for each boat ride" Stakeholders frequently contrasted their parents' livelihood strategies with the new livelihoods brought on by reef tourism. The potential to earn ten times as much as what their parents could earn in a week was a major topic in many interviews. Universally, people gave examples of behaviors that they used to practice but have since changed because they realized the links between tourism and reef health. "I used to spearfish on the reef all the time," said a dive industry worker. "I would never do that now. I have a reputation as an eco-dive shop. Tourists would not want that, my village would be angry with me."

Many stakeholders spoke of their children or the unborn generations that would follow them. Many invoked religion when discussing this topic, saying that good karma would be to leave their children healthy reef systems so that they too could stay in the village and benefit from the reefs instead of having to move into larger cities. "I do not want my children to have to move to Kuta. It is dirty there and filled with bad people. I want them to be able to earn a living in their home village, and the reef needs to be healthy for them to have this opportunity," said a tourism worker. "I am very lucky to have the job I do," said a waiter. "I do not think a lot about the reef because I cannot swim, but I do not throw cigarettes into the ocean or [use it to go to the bathroom] because I want to keep the reefs nice," he said.

5.4 Businesses That Promote Conservation

MPAs and local businesses that make their incomes from reef-based tourism both share some responsibility for conservation. Because businesses next to or within MPAs such as dive shops, hotels, resorts and restaurants benefit from reefs, one would expect them to play some role in conservation. This research found that co-managed MPAs had 73 percent of stakeholders who felt that businesses should invest in conservation, and 49 percent of stakeholders felt this way in Malaysian Marine Parks.

5.4.1 Malaysia: Out of gas

Of the businesses that did invest in conservation in Malaysian Marine Parks, most fell within the diving industry. Also, their investments tended to include ecological training for their employees, as well as debris removal from the reefs. Many dive industry respondents felt that their hands were tied by

Marine Parks and that partnering with them on conservation programs was not welcome. "It would make a lot of sense for Marine Parks to invite us into a program that would maybe clean the reefs, or let us set up a coral nursery or something like that, but it never happens," said a long-time industry worker. Many respondents agreed with this perception, citing two main reasons that Marine Parks does not invite businesses to invest in conservation programs or partnerships: territorial disputes and a lack of interest or will.

Business owners and stakeholders felt that Marine Parks blocked them from engaging in conservation. They tended to think that Marine Parks did not allow them to invest in conservation because Marine Parks lacked capacity to do so themselves. "You never see them out there," said one dive industry worker. "They know that we think this, so they do not allow us to do conservation projects because they think it will make them look bad." Another dive industry stakeholder with a background in marine science said that he asked to install a coral nursery offshore in the MPA and a small exhibit in the Marine Parks office on coral growing. "I was denied," he said. "Not sure why. I think they might just be petty." This idea of pettiness, territory and jealousy was mentioned by more than half of respondents in interviews and was a commonly cited reason for the inability of businesses to participate in conservation. Another stakeholder said, "They don't want us [dive shops] out there working on the reefs because we make them look bad. I never see them on the reef, all they do is stay in their office."

Marine Parks officers contested these opinions, however, saying that they did not ban businesses from investing in conservation but instead had a complex and rigid set of rules and permits regarding who can do what on the reef. "This is the point of good management," said a long-term Marine Parks officer. "We have many rules, and procedures. People must follow them. If anyone could do anything they wanted on the reef, who would stop the bad people from interfering on it?" Other Marine Parks respondents were skeptical of the aims of businesses looking to invest in conservation. "It does not seem to me that those are the right people to design projects for coral conservation in Marine Parks," said another Marine Parks officer. "In Malaysia, it is our custom that government manages valuable resources, and if we let businesses help, then what if they use things to their own advantage?"

Stakeholders from the dive industry frequently discussed how they removed large nets, piles of debris, televisions and other large items off the reef when the need arose. "Last month we removed a thousand pounds of aluminum siding," said one dive industry worker echoing the stories of many like her. "If we had filed a permit in Marine Parks, or asked permission, it would have taken months to hear back, and I bet they would have rejected it." Other dive industry stakeholders remembered times when they were asked directly by Marine Parks officers to remove a large piece of debris from the reef. "They told me they were low on petrol," said a shop owner in Tioman. "How does the Marine Park run low on petrol when they charge everyone who comes in so much money?" Another respondent on the Perhentian islands had a similar story: "The Marine Parks guys came to my shop and said they had seen a really long piece of fishing net on the main reef out there. They then told me that they didn't have enough petrol to go out and remove it and that their air tanks and dive equipment were broken." Another stakeholder from Tioman had a similar story: "The Marine Parks guys, when they do come in for work, come to me and are like 'we have no petrol' or 'our dive stuff is broken' and they want to use our stuff or they just want us to do it." Another dive industry worker quipped, "Marine Parks guys can never go get nets. We always go and do it because they're out of gas."

5.4.2 Indonesia: Put your money where your mouth is

Indonesian stakeholders on the other hand felt that businesses readily invested in conservation and valued the business owners in the community who were the most visible investors in conservation projects. A representative example is the owner of one of the largest resorts in Pemuteran, Pak Agung. Although he was not interviewed for this project, dozens of stakeholders spoke of his charitable work in the community promoting conservation and programs aimed at reef protection. Many felt that he was the first person in Pemuteran to promote the eco-conscious culture that the village has become famous for. According to respondents, Pak Agung bought his large resort in the early days of the MPA, and invested in the community so that people began to stop fishing the reefs and work in tourism instead. "His hotel was one of the first places that people saw and were like, 'hey working there would be a lot better than being a fisherman.' Many people echoed these beliefs, that the initial buyers of resorts in Pemuteran were major investors in conservation and the community itself.

It was because of these stakeholders that the livelihood transition was a peaceful one in Pemuteran, one, according to interviews, where people realized they could make significantly more money working in these early hotels versus fishing. "The thing that made these people different," said one local conservation worker, "was they did not go all out building way too many hotels. Pemuteran hotel owners were few and they were slow. They planned stuff, invested in the community, and worked for the reef." This same stakeholder compared Pemuteran to nearby Lovina. "Lovina did not get lucky the way we did, they were closer to Kuta, which made them explode with tourists. People bought up land and built too many hotels and a lot of the reef died, this did not happen here because businesses valued the reef and grew slowly." Another prominent stakeholder in Pemuteran called this phenomenon in Lovina the "Lonely Planet effect" after a series of books written for tourists. He had the following to say about development in Pemuteran and the links between businesses and conservation:

> This "Lonely Planet effect" happened to Lovina and not Pemuteran. Early hotel owners, like me, we wanted to do something called slow growth, where we would do development the right way. We knew the reefs here were unlike any others in Bali and we knew that if we let too many people come here too quickly, or if land prices shot up because everyone heard this place was the next big thing, then sedimentation would happen and the reefs would die. We grew slow, we invested in community groups, encouraged people to work with us for higher wages instead of dynamiting reefs, and the rest is history. We put our money where our mouth was, and we benefit from it today.

In co-managed sites, many respondents valued the contributions of local community members who owned businesses and invested in conservation.

Figure 5.2 An artificial reef structure placed near the reefs and financed by a hotel owner in Bali, Indonesia.

They were perceived as leaders even if they did not belong to formal MPA management organizations or had village leadership roles. Respondents frequently gave examples of ecological conservation programs funded entirely by local business owners. For example, one prominent conservation leader owned a dive resort, frequently installed artificial reefs in the bay for all tourists to enjoy (e.g., Figure 5.2), donated his boats for community events, hosted a children's traditional dance group and financed an NGO known as Reef Guardians who remove invasive species from the reef. Respondents viewed his work as an important part of the social fabric of the community and as having a large impact on the health of the reef ecosystem.

There was also the perception that business owners, while they were leaders and benefactors to the community, also owed these types of services to the community. "Business owners have a lot of money around here and they bring a lot to the community," said one dive industry worker, "but we also feel in the village that if you own a business that does well because of the reef that you should give back to the community because the reef is not yours alone."

5.5 MPAs Help Business

MPAs maintain the ecosystem while also by limiting certain activities on the reef that degrade the environment. But this activity actually helps business owners because healthier reefs bring more visitors. Stakeholders must be able to earn a living in the reef-based economy, with restrictions in place that ensure that the reef continues to survive. One important way that MPA managers support businesses is responding to crises such as unusual storm seasons with major damage to coral, bleaching episodes, disease outbreaks or illegal poaching, for example. These responses can be active, such as closing areas with enhanced thermal or disease-related stress so that visitor impacts from divers may not add to the stress. Responses may also be passive and include collecting data on bleaching or disease outbreaks. While passive responses may not immediately impact reef health, they may help in long-term planning for reef tourism operators who will be impacted by the effects of climate change. In Malaysia, 88 percent of respondents felt that MPA managers did not respond to crises in ways that allowed their business to thrive. In Indonesia, 81 percent of respondents felt that MPA managers did tackle crises in ways that supported business.

5.5.1 A tale of two mooring points

An example cited by respondents at both Indonesian and Malaysian sites of MPAs responding to crises included replacing mooring points. After a heavy storm season, mooring points come undone at many of the most popular reef sites. One stakeholder well versed in global conservation had the following to say about it:

I have been to Florida, where you are from. If the mooring buoy is not free, you go dive somewhere else. Not in Indonesia! If it is not free you drop your anchor, and some guys don't know better than to drop it on the reef. That's why it is so important to MPA managers to get out there, right when the monsoon ends and fix our mooring points. Most people here know better, but sometimes you have people coming from outside the village, and they don't care. They promised their tourists as day on a particular reef, and they will drop their anchor.

Malaysian MPAs share the same problem with Indonesian ones, where storm seasons undo many of the mooring sites (e.g., Figure 5.3), which is a big problem on popular reef sites. "Some of these sites get hundreds of small boats on them in a day," said one Marine Parks officer. "When they get washed away, we need to put new ones in fast because they are a major way that we prevent damage to the coral in the park."

Figure 5.3 A mooring point on Tioman Island.

Stakeholders in Amed, Lovina and Pemuteran all mentioned the importance of mooring points and how MPA managers need to install new ones in a
timely fashion. A majority of respondents noted that this does indeed happen.
"If it didn't happen, the dive industry would be at our door complaining,"
said one stakeholder in Pemuteran. He noted that the dive industry pays into
a fund every year in order to support the MPA managers in performing these
tasks. Generally, people perceived that stakeholders were happy to pay into the
fund as long as duties were consistently performed. People widely perceived
that they were performed and that the vigilance of MPA managers really did
support local businesses. "Even more than climate change," said one dive shop
owner jokingly, "anchor damage would do way more damage to my bottom
line. Before we had a system for local guys to come out and replace them,
people would drop anchors everywhere and break the coral." One of his
employees said, "I have cut anchor lines off the reef many times, but I haven't
had to do so in recent memory." When asked why he would cut those lines
in the past he said, "I saw them on big plating corals or big bommies, and
thought, 'hey if these get damaged, that is my paycheck!'"

Stakeholders in Lovina also noted the importance of replacing mooring
points and how in previous decades there was no system in place to finance
repairs after the storm season. Then, when MPA management organizations
became institutionalized through time, this lack of a system was replaced by
a solid system where villages took turns financing mooring point replacement
before every tourist season. "Before, guys would be like, 'I have no gas' or 'My
engine is broken can you guys do it?' and then we would not do it because it
was their turn and that wasn't fair. After we made rules, and a system, it is no
longer a problem." Repeated observations showed that the most popular reefs
around Anturan were in fact marked with buoys that were very popular with
local snorkel tour operators.

In Malaysia on the other hand, both field sites had stakeholders with strong
beliefs that Marine Parks did not respond to mooring point damage after major
storm seasons. Many thought that periods without mooring points would not
span single seasons but would instead continue for years at a time. "What
do they collect money from tourists for?" asked one dive shop owner on
Tioman. He continued, "The most shallow points are the worst. I see the
local guys running tours drop anchors right on the big plate corals out there."
This respondent was referring to the informal tourism operations that people
typically run out of their homes. They tend to include snorkeling and tend to
be cheaper than snorkeling with a dive shop in the MPA.

Other Malaysian business owners spoke of times when they had had enough
and opted to install new buoys themselves. One respondent said, "I had seen
the buoy missing for a whole season by Shark Point. That place is really

popular, and maybe 50 boats would use it a day. I complained a lot to Marine Parks, but finally I had to go and do it myself with my guys." Another respondent had a similar story:

After about a year of asking the Parks guys to replace a handful of mooring points where I saw people anchoring on the reefs, I just went into town and bought the cement and did it on my own. You know what happened the following year? They came by with their own bags of cement and asked me if my people could go and do some more for them. That's common with these guys. They don't do their job, and if you help, they expect it into the future. If anything, they hurt my business because they let their friends fish illegally, and then ignore problems throughout the park.

5.5.2 Fishing is banned. Or is it?

MPA managers' most basic responsibility is to prevent fishing in the no-fishing zones within the MPA. Without colorful fish, tourists would not come to snorkel and dive on the reefs. Additionally, both Malaysia and Indonesia have a history of dynamite and cyanide fishing, where the economically disadvantaged would use these destructive methods. Fishermen would use dynamite to stun larger fish on the reefs that are somewhat difficult to catch with a line and pole, such as snappers and groupers. Fishermen would use cyanide to stun small aquarium fish, collect them and allow them to recover and then sell them to the aquarium trade. In both Malaysian and Indonesian MPAs studied for this research, destructive fishing methods were widely regarded among all stakeholders to be things of the past but that occasionally would occur. The most basic task of MPA managers was to patrol for illegal fishing activity, be it destructive or otherwise.

On Indonesian sites, most respondents felt that enforcement was completely effective and that rule breaking was minimal among locals. "There is no way to break rules here," said one boatman in Amed. "Everyone knows everyone and you'll be taken off the reef in a minute, and forced to pay a fine." Likewise, in Lovina, respondents gave similar interviews: "It is impossible to break the rules on fishing here. We all live so close to the reef, I can look out my window and see if someone is breaking the rules." Similarly, in Pemuteran, a respondent had the following to say, "Our reefs have eyes on them at all times. The Pecelan Laut guys, NGO guys, dive shop guys. Everyone is watching, it is impossible to break the rules."

Stakeholders saw strong links between this enforcement and their business. "Having fish here is very important for us," said a small snorkel shop owner.

"Without the fish we would have no business." Another respondent said, "People in charge of the reef here do a good job of keeping people from fishing. It is important to me because people come here and spend money to see fish." Another respondent in Amed said, "Before locals began to manage reefs, many people would be fishing. Tourists, locals, everyone. Now that it is banned, many people come here and rent snorkel stuff from me, and it is good for business." Several respondents used to be fishermen, and shared these perceptions. "I used to fish," said a respondent in Lovina. "Then we decided that we would do tourism here and end fishing on the reefs. I make a lot more money now, and would stop anyone from fishing these reefs."

Even though many felt that the rules were impossible to break, I observed locals from the village spearfishing on the reefs a few times. This was a rare event and often happened at night. I began asking people about this informally. Respondents said that local people were sometimes allowed to break the no-fishing rules on the reefs. "You need to be from here through, and you need to only take a small amount, and you cannot bring foreigners," said one respondent in Pemuteran. I witnessed one man returning from spearfishing with several coral trout. When I asked him how he was allowed to he replied that "only locals can do it, and we may only take a few." He had planned to cook the trout for a small celebration that night in his new hotel. Off-the-record interviews with stakeholders in Amed revealed that the locals-only spearfishing rule was widely known among people who live there, but if visitors or expat residents were willing to pay enough, that they could definitely come along. Likewise, a restaurant owner in Pemuteran told me the same thing: "Here if you are a woman or if you pay a lot of money, you can break these rules." But he added, "You would need to do it quietly, and it does not happen very often." Indeed, my own observations showed no signs that spearfishing under the table was advertised anywhere in the case sites.

In Malaysian MPAs, most stakeholders felt that Marine Parks officers deserved credit for keeping illegal fishing off its reefs during the tourist season but not during the off-season. Several respondents spoke of how Marine Parks officers are widely thought to be absent from their job much of the time, but the lingering threat of enforcement keeps away much of the illegal activity that could otherwise occur on the reef. "I say a lot of bad stuff about those guys," said an NGO worker, "but I do think that their big, scary looking boats do prevent some people from fishing the reefs." Another respondent agreed and said, "I criticize Marine Parks for everything. But I would rather have people think they are here so they do not fish. Stopping fishing is really important for the dive industry." Many respondents shared the view that the Marine Parks fishing program had some links to the success of their business.

This was balanced out by the strong views that many held about the lack of fishing enforcement in the off-season. Many respondents discussed how as soon as tourists would leave and resorts would close, the fishing boats would come in and Marine Parks would not stop them. Many respondents felt that this happened because the people who owned the trawlers that would violate the bans were well connected and rich. "Marine Parks officers do well, they work for government," explained one respondent, "but they can add a large bonus in the form of bribes from fishing companies in the off-season." The idea that payoffs were behind the illegal activity was widespread among field sites. Marine Parks officers blamed the lack of serious enforcement in the off-season on bad weather and limited resources. One Marine Parks officer noted, "The storms are intense during this time. We miss illegal fishermen from time to time because patrols are not possible during big storms." Another respondent who worked for Marine Parks agreed and said, "We do not have any visitors paying entrance fees in the off-season, so we have a smaller budget for enforcing fishing bans. Plus, the weather may keep us on dry land."

Many in the dive industry noted that they had seen dozens of examples of tourists buying garbage bags of fish from fishing boats parked just outside of the MPA. "Marine Parks doesn't stop this because they get fish out of it," said one dive shop owner. The line of vessels just outside the MPA boundaries could be seen from the shore, especially on the small Perhentian Island. I witnessed one example of this garbage bag trade in fish and asked the people who had it where they bought it from. They replied, much like interviews had suggested, that it was from the line of boats just outside of the MPA. They did tell me that these fish were caught outside of the MPA. Though some of the fish could have been caught outside of the MPA, there were several reef fish in the garbage bag. "This is not uncommon," said a conservation worker in the MPA. "These boats are skirting the line all the time and catching fish on the reef, even during tourist season. I pull their nets off the reef all the time."

5.6 The Role of Civil Society

NGOs can act as a support system for MPAs and communities that depend on reefs for their livelihood. This study found that stakeholders in both Malaysia and Indonesia viewed NGOs as a critical component of their ability to make a living off the coral reefs.

5.6.1 NGOs and scientific monitoring

Indonesian and Malaysian case sites both had problems with a lack of scientific monitoring. All case sites lacked a permanent monitoring plan where water quality, thermal events and disease outbreaks could be monitored. In both Indonesian and Malaysian MPAs, the closest thing to organized monitoring was the ad hoc efforts of NGOs. For example, in Amed, I witnessed a multi-day training for dive industry workers on identifying bleaching corals and loading this data into a single database. According to several respondents, programs like these are spreading across Indonesia and are actively focusing on scaling up or improving ways to collect more data and apply it to management. One weakness with the focus on bleaching was that many respondents who had participated in the training were frustrated by an inability to act locally to prevent bleaching. "We are good at protecting local reefs, but even though we try, we cannot do anything when bleaching happens. Collecting data is important for scientists, but for us locals, I am not sure I understand." The view that programs that support stakeholders collecting ecological data had confusing aims was widespread among interviews. Another respondent said, "They have us collect this stuff, and others have trained us to collect data on disease, but I am not really sure why. We cannot change our behavior and stop bleaching." It seems that more education explaining data collection is needed. Despite the widespread view that many locals do not understand the point of programs because individual actions cannot end environmental phenomena such as bleaching, ecological monitoring is still important.

Malaysian MPAs also had major efforts for ecological monitoring conducted by NGOs, primarily Reef Check Malaysia and its biological surveys in Malaysian MPAs. These surveys include data on biodiversity, coral cover, fish counts and rugosity. "The Reef Check Malaysia data is very important," said a conservation stakeholder based at a local university. "They've been collecting it now for ten years, so soon we will be able to talk about trends on Malaysian reefs thanks to their work." Another stakeholder said, "It is really unfortunate that Marine Parks does not have its own ecological monitoring program. But, we have Reef Check so it is okay. I worry that Marine Parks guys don't use Reef Check data however." This was a source of confusion among respondents much like in Indonesia. Stakeholders were unclear over whether management decisions were made using Reef Check's data and the scientific work of other NGOs. One stakeholder said, "I worry that Marine Parks officers have the data, but lack the capacity to apply the findings, or just have no central spot to organize it." Stakeholders were generally concerned about a lack of scientific literacy among Marine Parks employees, arguing that even if they have access to the data collected by NGOs that they might not know what to do with it.

Marine Parks employees discussed their own agency's focus on management and not on science. "We work with collaborators for the science part," said one Marine Parks officer, noting that "building collaborations with the community is important and it helps many people be involved with the work that we do in Marine Parks." Another Marine Parks employee discussed the fact that science was conducted by NGOs and university partners positively as well, saying, "We involve the local community by asking them for help with science. It benefits the university and it benefits us."

5.6.2 NGOs link communities and ecology

In both Indonesian and Malaysian field sites, respondents discussed how NGOs engaged the community, bringing ecological monitoring capacity or knowledge to local residents. For example, the Reef Guardians in Pemuteran have a beach front office on Pemuteran Bay that educated locals and visitors on the impacts of invasive species and coral disease. A major part of the program is financing dive certification for local villagers and training them to collect invasive species off the reef that also damage coral. "Without this training, I would not know how to dive, and I would not know how to remove crown of thorns starfish," said one of the Reef Guardians volunteers. Others in the community spoke of the spread of ecological knowledge caused by Reef Guardians and other similar NGOs: "Without the Reef Guardians, the danger of crown of thorns starfish destroying the reef would not be widely known. I see dive masters removing them from the reef when out on dives, so the education is working." Other local NGOs finance efforts aimed at encouraging locals to behave in certain ways that positively impact the environment. For example, for every endangered turtle nest that locals leave intact and report, a local NGO pays them. "They payment allows us to not only track the locations of nests and mark them so that beach goers won't trample them, but also gives locals a reason not to dig up nests and eat the eggs." Another respondent commented on how this NGO began to spread a new social norm of leaving turtle nests in place:

The Turtle Project in Pemuteran had big effects on the community. The first was that before, people did not know that turtles were endangered here, which is why people would dig up the nests and eat the eggs. Second, people did not know that if you dug up the nests, that tourists would be upset by that. The NGO promoted behavior where we left nests in place, and showed tourists how this community protects turtles. Also, it helped people realize that people would be willing to pay to come to a place that protects turtles, and the really naughty people learned this by the NGO actually paying them not to harass the nests. The other benefit was that the NGO kept a record on how many nests we had each year, what turtles nest on local beaches, and where they were on the beach. So they are monitoring turtle population trends and ecology in our community.

Similarly, in Malaysia, respondents discussed the importance of NGOs that engaged the community and expand educational opportunities or capacities relating to ecology. Blue Temple Conservation on the Perhentian Islands, for example, focused on youth outreach to children and teenagers, educating them on marine species and environments. They focused on raising awareness of the importance of marine life as it related to life in the village. Several employees of Blue Temple spoke of the lack of widespread knowledge in the community that linked a healthy marine environment to the health of the nearby village. Focusing on children and engaging them with fun activities such as art projects focused on marine species identification allowed for village kids to begin to build a relationship with local ecological systems and to value these systems more than previous generations. "The understanding that the health of the reef and the village are linked is low among the older generations," said one employee. "The youth programs aim to fix that." Interviews with many in the village suggested that the presence of Blue Temple had changed their mind on the importance of the marine environment. "My children spend a lot of time at the Blue Temple clubhouse," said one woman. "They come home and teach me what they learn. We are learning together."

5.7 Summary and Conclusions

I found that co-managed institutions were more successful when it came to integrated management where the MPA, through its conservation activities, helped households earn their livelihoods by making rules that did not prevent people from earning a living, and if they did, they offered support for livelihood transition such as requiring dive shops to hire locals. Additionally, evidence of businesses ranging from hotels to restaurants to dive shops engaging in some form of conservation was widespread. Businesses that depend on reef tourism performed reef cleanups voluntarily, built artificial reefs and intentionally hired locals who would otherwise have to fish. Hotels offered options for reducing water, energy and plastic waste. Restaurants offered information on where their fish were caught. MPA managers and village leadership required dive tourism or general tourism stakeholders to hire a certain proportion of local people, which solved the issue in the early days of how fishermen would transition away from reef fisheries.

On the other hand, there was a real trend among Malaysian stakeholders, even those who worked directly on dive boats or water taxis, where they believed conservation was not important. Many interviews suggested that education and low incomes could be an explanatory factor in these perceptions, but that is a story told all too often. I would instead focus on Malaysian civic culture, where the government is in charge of environmental management, as the main explanation for the perception that conservation and reef tourism livelihoods are not linked. Malaysians see any and all environmental work, from enforcement on the reef to litter removal to reef biology as the government's responsibility. This has created a mindset where individual behaviors, such as leaving litter on the beach after a meal, are an all-too-common occurrence in Malaysian MPAs. The irresponsible nature of this behavior does not register because the government is there, looming large in the minds of locals, and will fix the problem of litter on the beach. Unfortunately, when locals believe the government to be large and all-powerful—and in fact it is quite the opposite—a vacuum of responsibility emerges even in an official MPA. Additionally, tourism stakeholders had no requirements to hire local people from the village who were forced to stop fishing the reef when the MPA was formed in the mid-1990s. Many migrant laborers from Indonesia and Bangladesh constitute the majority of the hotel and restaurant workforce on both Tioman and the Perhentian Islands. Thus, hotels and restaurants benefitted from reef tourism, rapidly moved in and continued to develop; previous generations of fishermen lost their livelihoods and had few alternatives; few alternatives exist today for their children who often move to larger cities to find work. Many local villagers in Malaysian MPAs view conservation or behaviors that contribute to conservation as helping people who are already rich or helping foreigners continue to come in and work. They do not see their personal economic well-being as linked to the reef because the MPA managers, combined with local businesses and their hiring practices, have never gave them a reason to.

In Indonesia on the other hand, their civic culture born from the legacy of the corrupt Suharto New Order dictatorship created a widely held perception that government is ineffective, a perception that has spanned the past three decades. Even though power emanated from Jakarta, Indonesian graft and bureaucratic ineffectiveness was a fact of everyday life and, thus, locals had a lot of de facto control even before it was official. Thus, the mental transition required by locals to take responsibility over their resources was already in place, and can probably explain why people so consistently link local conservation behavior and livelihoods. That is not to say there is not work to be done. The litter problem in Bali is legendary. It is the first thing that any taxi driver talks to you about, and it is one of the most common topics of conversation among tourists and locals when discussing the environment. Explanations for the litter problem can include the fact that plastics and Styrofoam have only been in use for 40 years on the island where, prior to that, banana leaf was used exclusively for packaging. Thus, there remains a challenge connecting local-level behaviors, such as littering, and livelihoods in reef tourism, though the perception that the two are unrelated does not exist in Bali as severely as it does in Malaysia.

I also found that both top-down and bottom-up MPAs need NGOs to fill the capacity gaps in scientific monitoring. Reef Check Malaysia, Reef Check Indonesia and Conservation International's Bali offices were the major sources of scientific monitoring support. Both Reef Check and Conservation International are global NGOs, meaning their funding comes from all multilateral institutions and donors over the globe. The local staff is remarkably dedicated not only to reef conservation but also to the difficult task of including and empowering communities to protect their reefs. Locals in Indonesia and Malaysia noticed their dedication and had overwhelmingly positive things to say about the employees of these organizations and the programs that they started and continued. NGO workers were widely known in the village, and were widely trusted as a source of action when problems befell the reef. In addition, when people did not know how to answer my questions they would insist I talk to a certain person from one of these NGOs, and often, locals had their phone number in their mobile. I would argue that this accessibility, openness and trust that a few dedicated NGOs have built in communities over the years helps them be successful in launching and continuing effective conservation programs.

International NGOs were not the only source of scientific capacity-building in the villages. Local NGOs, started by a near-even mix of Southeast Asians and expatriates also filled this role. The Perhentian Islands, for example, has a few local NGOs funded through volunteers who travel from their home country in order to come to the island and do marine conservation work, which can range from rubbish removal to biological surveys of reefs. Again, it is the leadership qualities and dedication of their founders that make these NGOs so effective and respected in their village. Some of them single-handedly educate village stakeholders on marine issues, and much of the scientific work they do would not otherwise be done.

I also found that Malaysian Marine Parks must allow NGOs more flexibility to engage in conservation, even if it is primarily in community outreach and education, namely educating locals on the links between reef health and the economy. Although Marine Parks offers programs for schools and limited outreach for the community, it does not have the desired impacts where locals internalize the importance of individual behavior, conservation and long-term economic viability of reef tourism in the Malaysian Marine Parks. This is where the local level-NGOs are having a high-level of impact. They are teaching villagers links between the use of plastic, reef pollution and diminishing tourism. They show villagers how littering the beach during holidays means that tourists will not want to come, and when tourists do not want to come, incomes decrease.

In the opening sections, I introduced two hypothetical MPAs, Case A and Case B, and described what integrated management looked like in Case A, and what non-integrated management looked like in Case B. I also argued that integrated management requires balancing the needs of different stakeholder groups and limiting the way people can access the reef, which invariably creates winners and loses. Once this happens, the MPA institution as well as the stakeholders who got what they wanted can combine efforts in order to ameliorate the status of the stakeholders who did not. The Balinese cases demonstrated the balance between different stakeholder groups. In the early 1990s, when the reefs were closed to fishermen, dive sector entrepreneurs made efforts to hire fishermen and, likewise, fishermen made efforts to transition into the dive industry seeing the potential to earn a greater income compared to fishing. Comparatively, in the Malaysian MPA sites, fishing was banned before there was even an active dive industry in place, so that there were no winners and no losers, only losers without a second industry that they could transition to. Once the Malaysian Marine Parks built offices on the islands and dive tourism began to take over as the dominant economic sector, long-standing resentments remained and fishing bans in the MPA became something that is only enforceable for part of the year.

Compared to Bali, where many fishing industry stakeholders became diving industry workers, Malaysia saw a pause in livelihood transition that can also explain why stakeholders do not link conservation and livelihood. In Bali, the common logic that underpins local stakeholders' support for conservation is that while previous generations were fishermen, that was a hard living; now tourism is more profitable, safer and easier than spending weeks at sea. Also, while previous generations removed fish, the new generations leave them in the water and earn a living this way. This mindset is not there in the former fishing villages within Malaysian MPAs, probably because there was no alternative livelihood when the fishing bans were put in place and the MPAs were formed in the mid-1990s.

The final ingredient for integrated management that balances economic and ecological needs is the necessity of the steam release. MPAs in Indonesia, for example, allow line and pole fishing by people who were born in the village, with a visible limit that can be monitored by the numerous people on shore. The idea of steam release does not apply in Malaysian MPAs because Marine Parks Officers do not enforce the fishing bans in a serious way. Perhaps if they were to enhance their enforcement, locals would be allowed the use of line and pole and small bag limits on catch.

Chapter 6

LEGITIMATE MARINE PROTECTED AREAS

6.1 Overview: Stakeholder Perceptions on Legitimacy

In this chapter I discuss the different perceptions on marine protected area (MPA) legitimacy across Indonesian and Malaysian sites. Legitimacy is when resource management decisions are shown to accomplish stated objectives, or where management decisions are made in *appropriate ways* (March and Olsen 1989). An institution must have the "right to rule," secure compliance and issue sanctions for non-compliance (Buchanan and Keohane 2006, 3). In order to operationalize this definition in the field, I gauged legitimacy in three ways by asking stakeholders (1) whether they felt the organization actually does its job and effectively manages the reef, (2) whether the MPA is something they value in their community and (3) whether power-sharing between government, MPA managers and local stakeholders occurs.

Stakeholders in Indonesia and Malaysia were asked three questions to measure legitimacy. The first question asked whether decisions on reef management accomplish objectives and whether they were made in appropriate ways by the MPA. The second question measured the "right to rule" of the MPA by asking stakeholders whether they value it. The third question, different in Malaysia and Indonesia, asked stakeholders whether they think that power sharing arrangements are in place, and whether these are arrangements that increase stakeholder compliance with rules. During the survey and interview process, stakeholders from both sites added additional components to my own definition of legitimacy. These included the following: (1) the use of science in decision making, (2) the visibility of MPA leaders and their interaction with local people and (3) leaders and members of the MPA taking responsibility for their role as reef protectors in the village.

As in the previous chapter, I offer the example of two hypothetical MPAs, Case A and Case B in order to demonstrate what legitimate MPAs look like, Case A where there is legitimate management and Case B where there is not. Beginning with the MPA in Case A, when stakeholders speak of their reef management institution, there is a widely held perception that the dozen or so men who make up the MPA's members are always "on the job." They are regularly seen patrolling the reef by boat, or they regularly report illegal activity to the police or head of the village. When local people are asked about the MPA, they say that they appreciate it and respect those who join. Some even discuss moments where they have broken rules and have been punished by the MPA with small financial penalties. However, they still acknowledge that the MPA is important and the sanctions showed them that it is better to not break the rules. They link the importance of the men who work in the MPA to the health of the reef and they appreciate the time that MPA members dedicate to their jobs. Local people feel that the MPA is "in charge" and that the federal government or even the state or provincial government trusts them to look after their own reef. Because of this responsibility, locals have increased pride in their own MPA. People with a wide range of educational backgrounds believe that the rules are fair because they are based on science. They also know that scientists are monitoring their reefs, and if rules are broken, the reef will become degraded.

The MPA in Case B is not legitimate. Stakeholders widely speak of absenteeism among those who are supposed to be making and enforcing rules on the MPA. People speak of the fact that rules exist but that nobody is there to enforce them. Many believe that those who work as members or employees of the MPA make too much money and do too little work. Stakeholders speak frequently of the tricks that MPA workers use to avoid having to do their job, including signing into the office and leaving immediately after. Not many people from the local village work for the MPA; often it is outsiders who get hired to perform these roles, and many times it is people who lack knowledge of marine science. The reason they take these roles is that the pay is generous, and there is the possibility that after working for the MPA they can secure another official role with a few years of experience. Patrols of the reef are rare. For that reason, many locals do not value the MPA. People discuss times when they have seen blatant rule breaking, such as large trawlers operating in MPA waters that go unpunished. Others describe token gestures of enforcement, where illegal fishing boats may be stopped on one day but are back out on the water the next day. People do not see links between the MPA, the people who staff it and the health of the reef. Instead, they think that behavior on the reef is a bit of a free-for-all, where "anything goes" because of a lack of enforcement. Many do not see the rules that govern the MPA as based on science because there are no scientists studying their reefs. The following results and subsequent discussion sheds light on how Malaysian and Indonesian MPAs contain aspects of legitimacy of these hypothetical cases.

6.2 Survey Results

Figure 6.1 show the statistical tests that I completed on three questions meant to query stakeholders on legitimacy.

6.3 Different Perceptions of Institutional Efficacy

This study found that while 23 percent of Malaysian respondents said their reef management institution protects the reef, 74 percent said that Malaysian

Does the reef management organization actually protect the reef?

	No	Yes	Total
Malaysia	74 (0.77)	21 (0.23)	95
Indonesia	49 (0.32)	104 (0.68)	153

$$p = 0.000{*}{*}{*}, z = -7.02$$

Is the reef management institution something that you value and respect or does it not work?

	No	Yes	Total
Malaysia	93 (0.86)	15 (0.14)	108
Indonesia	38 (0.32)	120 (0.76)	158

$$p = 0.000{*}{*}{*}, z = -9.94$$

Are there official power-sharing arrangements that devolve responsibility to the village?

	No	Yes	Total
Indonesia	13 (0.11)	110 (0.89)	123

Do nongovernmental organizations (NGOs), either local ones or international ones, support you in conservation?

	No	Yes	Total
Malaysia	58 (0.76)	18 (0.24)	76

$$p = 0.000{*}{*}{*}, z = -9.41$$

Figure 6.1 Survey responses to questions on legitimacy. *** : $p < 0.001$.

Marine Parks does not protect the reef ($p = 0.000$). Of the stakeholders who did say that the Malaysian Marine Parks protects the reefs, 70 percent of those respondents were Malaysian Marine Parks employees. If local stakeholders in Malaysian MPAs do not believe the MPA protects the reef, then what do local resource user stakeholders perceive that Marine Parks officers do?

6.3.1 Malaysia: Invisible maintenance and park facilities

One dive shop owner on the Perhentian Islands who co-owns a long-running, successful shop and is widely perceived as a local conservation leader said,

> I never see those guys [Marine Parks officers]. When I do see them, they are parking their fancy boat in front of our place, and they come into town so one of the hotels or restaurants gives them a free lunch or a free sandwich. I only see those guys when they are hungry.

Many interview respondents shared the perception that ongoing mainte-nance linked to reef conservation does not occur in Malaysian MPAs. For instance, a long-term dive shop owner in the area outlined a series of practical, day-to-day duties that Marine Parks officers should complete, which would include improving floating buoy barricade systems that separate coral areas from areas where swimming is allowed. Additionally, there are dozens of plastic floating platforms (Figure 6.2) across the marine park for recreational use that come unhinged from their anchor and need to be located and reattached. Last, there are mooring points for a range of recreational dive

Figure 6.2 Floating platforms. Left: Gray floating platform installed near patchy coral reefs by Malaysian Marine Parks. Right: Plastic floating platforms on Tioman Island MPA.

Figure 6.3 Dive industry workers in the Perhentian Islands MPA.

and snorkel vessels to attach when visiting reef sites popular with tourists (Figure 6.3).

When I asked respondents how frequently Marine Parks officers fulfill these duties, a common response was that it happens very rarely if ever. One dive operator in the area had the following to say about these types of responsibilities for marine parks officers:

> One time after a big monsoon season, quite a few plastic platforms near a really popular reef came undone. For the first few weeks of the season, you'd see the people all over the corals, stomping on them, kicking them, everything. I went to Marine Parks offices each day trying to ask when someone will be out with the cement and equipment to reattach then.

One of the head guys eventually came down to see me after weeks of nagging mid-way through the season and tried to hand me the cement, line, and buoy asking if my dive instructors who work for me can go out and do it. "We don't have the people or the time to do it" was their take on the matter. I asked him how I am supposed to lose a day's work in order to do their job for them.

Another common observation was the perceived lack of MPA staff performing necessary functions while continuing to collect a 20 Malaysian Ringgit tariff (approximately $5) from every visitor to the park. A near universal opinion among stakeholders was that given the volume of visitors who come into the park each year and that every visitor must pay this entrance fee, the lack of capacity, time or resources to perform this type of small maintenance on the park system was inexcusable. Stakeholders roundly criticized Marine Parks officers for asking local dive shops to pick up slack and repair mooring points or floating platforms. In their view, it was unfairly passing off costs, time and labor onto private businesses while pocketing the visitor fee. Common words used to describe Malaysian MPAs included invisible, absent and ineffective.

6.3.2 Malaysia: "Where is the science?"

A second concern voiced among stakeholders was the "lack of science" being conducted within the parks. Many described a mismatch between what the duty of the Malaysian Marine Parks should be: science, conservation and management, compared to what it actually was on the ground. One stakeholder who had worked for decades on Tioman had the following to say:

> I have been here since I was a kid. When I was a kid, we used to swim down and see who could fit inside empty giant clams; that was how big they were at the time. Nowadays, giant clams, or the only ones that are left, are no bigger than your head. You know why that is? Because there are no scientists on staff at Malaysian Marine Parks. Where is the science? Not a single biologist, no environmentally trained people. Just guys who sit up in that building and do god knows what, collect big paychecks from government. I'd do it too, anyone would, the pay is good, the perks are good, but look at the reef. You can see the consequences.

A widely held perception among informants was that Marine Parks was nothing more than a large, flashy, expensive building where little science and little work goes on. One Marine Parks officer who asked to be referred to by the pseudonym Ali, gave an extensive interview responding to much of the criticism outlined above. Ali outlined his educational background and what he was trained as in a field vastly different from any type of environmental conservation.[1] He explained that government jobs in Malaysia are distributed in a way where you cannot often control what specific bureaucracy you land in, but the pay, benefits and social prestige are so highly regarded that they outweigh the fact that you may have to be stationed on an island in order to receive them. Ali candidly explained that many Marine Parks officers are in fact biding their time and "putting in their dues" in hopes that they will be transferred to a different bureaucratic entity after a few years of experience. Ali did note, however, that at times, Marine Parks does hire locals from nearby villages who have conservation-oriented educations and want to work actively to better their local reef resources, but this was the exception rather than the rule. Ali commented on the lack of scientific research being done in the parks on the part of the Marine Parks themselves as a positive thing:

There are many criticisms that can be leveled against Marine Parks, but I do not think that the lack of scientists on staff is a very good one. It's not like there are no scientists at all, instead we emphasize making partnerships with local universities and allow them through permits and sometimes funding and assistance to conduct their own research in the parks. This helps raise the prestige of Malaysian researchers, and it helps to outsource some of the functions of the Marine Parks so we can focus on visitor education and maintenance. Our big function here is management, which means repairing the buoys and recreational areas, not doing biology or fisheries science.

Other stakeholders criticized what they said was an outdated idea of "management" in the Marine Parks. For example, when I read the above comments to a well-recognized conservation leader in the Malaysian Marine Parks system, he told me that this was a widely recognized problem in Malaysia's NGO community, whereby Marine Parks officers see themselves as "there to put floating platforms and swimming areas in place," instead of "managing" the reef resource. One example given by many informants of a type of management that should be used more is strategic area closures. This was a widely held idea among stakeholders when asked the follow-up question to interview, "If you had unlimited money and time, what would you do as a Marine Parks officer or ask the Marine Parks officers to do?" Many replied that when monsoon season comes through, some of the highly used reefs

---

1 Detail and duty station withheld to preserve anonymity.

receive heavy damage. Especially if the summer happens to be characterized by exceptionally high temperatures and bleaching, further weakening the reef system. Many respondents suggested that heavily damaged areas might need to be cordoned off for some time. Many stakeholders commented on how area closures were a cheap, easy and a popular intervention to the system. Many remarked that since the Marine Parks were so small that violators could be seen from shore, Marine Parks officers have the boats, staff and time on their hands to do regular patrols to enforce that they are a low friction management intervention.

Some pushed back on this idea, however, saying that there was a difference between the dive industry and the budget operations that characterize poorer households in the village. In each of the Malaysian Marine Park sites visited for this dissertation, there was roughly a dozen dive operators on each of the islands, with well-crafted publicity operations that include things like websites and online booking systems. There were also businesses more along the lines of a mom-and-pop operation, usually run out of the back of a private home, with rudimentary signage, no online presence, a focus on snorkeling and not diving and limited English. They were marketed to domestic tourists, and had a defining characteristic of smaller boats compared to the dive operators, as well as a vast majority of tourists on board wearing life jackets for the duration of their tours because of their inability to swim. These smaller businesses were cheaper. In interviews, when I asked about the need for periodic closures, these stakeholders said that they were against them because they would get in the way of business. They did not believe that occasional closures could, in the long term, lead to healthier reefs and better business. There was a suspicion of the dive industry, whereby some of these informants suggested that if the Marine Parks were to close certain dive sites, the dive operators who market to Western tourists may still be able to access them, and the smaller snorkel tour operators would be shut out and punished.

6.3.3 Indonesia: MPAs "get the job done"

The perceptions of stakeholders in the three Balinese villages vary drastically when compared to stakeholders in the Malaysian Marine Parks. When asked whether or not they believe the MPA manages the reefs, 68 percent said that they do. The words of one boatman in the village of Kalibukbuk, Lovina, illustrate what a majority of respondents had to say regarding management:

> Of course our reef management cooperative manages the reef, nobody else would do it if they didn't do it. Local government doesn't know how, and it is our responsibility by law. But it is also our livelihood. If we did

not look after our reefs, we would have no tourists and no money. We get the job done.

Stakeholders widely held the perception that local people who manage the MPAs perform reef conservation. Take, for example, the village of Anturan, Lovina. There, the elected head of the MPA that oversaw the local dolphin tours, snorkeling and dive operations in that village, spent over an hour translating his organizations rule book into English during our interview and explaining the significance behind the rules. The most important rules included pricing schemes for tours, informing tourists who rent snorkel equipment not to touch corals and anchoring only on designated mooring points. These prices were set in monthly meetings and fluctuated based on changes in gas prices or unexpected increases or decreases to the numbers of tourists visiting the village. He explained that fixing prices was absolutely critical to securing the legitimacy of the organization and the trust among its members. "Without price fixing rules," he explained, "people back stab one another, and there is a general climate of mistrust which leads people to do bad things to one another, and abuse the environment."

Figure 6.4 MPA members dragging their boat to its assigned spot on the beach.

MPA leaders and members alike praised their system of entry whereby everyone was assured at least one boat of customers every week. These rules allowed stakeholders to create what they perceived as a fair, limited, and organized access system to the reef, called *antrean* in local parlance (e.g., Figures 6.5 and 6.6). This system ensured that that all of the village's boats would not be on the reef or performing marine mammal tours simultaneously, and instead, a rigid turn-taking system was in place so as not to overcrowd. This ensured that boat captains would not steal business from other MPA members and also that tourists would not feel as if locals overcrowded the reefs. MPAs put graduated sanctions in place to ensure rule breaking did not occur. This distributed customers equally among the villagers, and in conjunction with price fixing, resulted in a scenario where conflict is rare. Stakeholders viewed the MPA as a source of stability that was critical to the well-being of everyone working in reef tourism, frequently using the following words to describe it: stability, relationships and fair.

Another major source of legitimacy were the monthly meetings in the MPAs of the villages. These were forums where problems could be discussed, solutions could be posed and action could be voted on. One example of a meeting that informants discussed was when a popular mooring point was destroyed by a big storm and a meeting was called between two reef management organizations in two neighboring villages in Lovina in order to engage in a type of cost-sharing and labor division to reinstall the mooring point. One stakeholder reflected on the importance of these meetings:

Without these meetings, there would be a lot of conflict. At these meetings, you can get everything out in the open so your heart is not heavy with anger at other men who you work with. We meet like a family, we settle things, we punish rule breaking. I was punished once for asking for higher prices and you know what they did to me? It is called "scorcing" it means the leaders take your boat and impound it for one month, I was very hungry that month!

This particular reference to the graduated sanctions in this particular village, known as "scorcing" (pronounced *score-sing*) was echoed by nearly everyone interviewed. Everyone had been scorced at least once, and they would frequently laugh when recounting their mistake and the accompanying month of hardship that would follow when they could not work because their boat was impounded. The good humor was a reflection of the buy-in to the level of punishment that scorcing embodied and a nearly universally held belief that even though it was a true hardship at the time, ultimately everyone benefits when rules are followed.

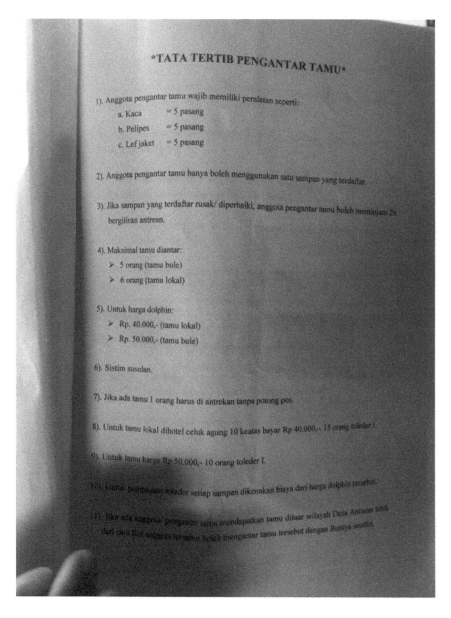

Figure 6.5 Rules for members of a co-managed MPA in Lovina for men who make a living bringing tourists on either dolphin tours or snorkel trips. Rules include how much equipment needs to be in boats (e.g., "5 life jackets"), how many people are allowed in boats (e.g., "6 locals or 5 foreigners"), prices for dolphin tours, special pricing for guests of local hotels and other practical guidelines.

Figure 6.6 An MPA leader in Anturan village explaining the rules of the institution.

Stakeholders across Indonesian sites held the perception that their MPA is the main source of reef protection. Common words used to describe these ideas included *strong* and *our* reefs, in addition to words that conveyed necessity, such as we *must* and we *have to*. Stakeholders felt that the Indonesian government across scales lacked the capacity, know-how and proper anti-corruption measures to perform management. Many expressed a mix of pride and relief when discussing the policies that devolved reef management to the village level, explaining that devolution meant that something that was already happening became legal. Many stakeholders contrasted the capacity of MPAs to manage with government's inefficiencies. Take the following excerpt from a conversation with a one-time dynamite fisherman who had since reformed his ways and now works in dive tourism in the Lovina area:

Before the villages were in charge of their own reefs, anybody could do anything. You want to go spearfishing on the main reef? Okay. Especially if you're a tourist from the West. You want fried sea turtle? Fine! There were absolutely no rules here. Through time, as villages realized they were in charge, key people rose up and really took control. We go to the same temple, we are connected in my village through family and through religion. These connections and the changes where the village was put in charge of our local environment converted me from a dynamite fisherman in the 1970s to a dive tourism worker in the 1990s. We have to make changes like these because the reef is ours.

6.3.4 Indonesia: All for show?

Although the majority of stakeholders perceived that Indonesian MPAs are in charge of managing the reef, there were important dissenting voices. Members who disagreed tended to cite membership as a source of legitimacy, arguing that when membership is not required among all resource users, its activities involve less management and more socializing. I focus on the case of Pemuteran where such perceptions were the most widespread. Pemuteran is a tourist destination gaining global recognition in the form of awards from prestigious multilateral development organizations such as the United Nations Development Programme (UNDP) for its efforts to promote sustainable tourism. A minority, albeit a sizeable minority, of stakeholders disagreed in interviews that the MPA members were actually managing the reefs. They gave three reasons for this perception: First, unlike other nearby villages, membership in the reef management institution in Pemuteran was voluntary as opposed to a required responsibility of those who use the reef. As a result, it had more of a resemblance to a social club than an MPA and was therefore not a legitimate management institution. Second, much of the day-to-day activities of the organization, such as weekly reef patrols, were more "for show" as one respondent put it than they were for management purposes. Third, there was a pervasive lack of science occurring in the management of local reefs.

Similar studies lend credence to the voices of these stakeholders. One of the most important institutional characteristics of successful versus non-successful MPAs involves rules that govern membership. These rules specify how membership is achieved, delineate who is required to have membership and help avoid a free-rider problem within the reef management institution (Cinner et al. 2009b; Cinner et al. 2012; Stevenson and Tissot 2014). If membership is mandatory for all reef users, problems of free ridership are solved and institutional theorists hypothesize healthier reefs. If membership is optional and criteria to join is vague, reefs of poorer quality were expected.

To characterize membership rules of Pemuteran's MPA as simply informal would be misleading. The *Pecelan Laut*, roughly translating to "Hindu sea

Figure 6.7 Pemuteran Pecelan Laut. Top: Pemuteran Pecelan Laut in their office. Bottom left: The office. Bottom right: Meeting area and office.

police," or those who manage the MPA, are best described as a social club, religious group and highly esteemed village organization (Figure 6.7). The high amount of admiration directed towards the Pecelan Laut meant that many had strong desires to join. Respondents repeatedly said things like "I will join the Pecelan Laut one day if I work hard" or "It would be important to my family if I could join the Pecelan Laut," or spoke of joining after monumental life events such as purchasing their first business. This suggests that mandatory rules that compel reef users to join an MPA may not be necessary if the community holds the institution in high regard, as it does in Pemuteran.

Those who felt the Pecelan Laut were "for show" argued that actual decision-making authority was held by tourism elites (TEs), defined here as hotel or resort owners who have invested in at least one high profile conservation effort in the village. TEs are visible figures in the village, well known to all local stakeholders, as their businesses are major sources of local employment. TEs are known in Pemuteran for implementing conservation efforts, funding enforcement, spearheading restoration programs and behind-the-scenes financial payoffs of village leadership and local politicians that result in conservation. Not all TEs were characterized favorably. Respondents spoke frequently of the need to be "well connected" to "important people" in order to have the capital that would allow a local to buy a hotel. One respondent's words illustrated this perception:

In Pemuteran we got lucky. Our hotel owners care about the environment. But this is Indonesia. Someone could have moved in years ago, someone who knows powerful people, and they could have destroyed all the reefs to build a massive resort. We got lucky that those who came in, in the early days, had money and care about the reefs. But money means that these people have a lot of power here, and are in charge of many things. They make decisions around here, and some of them do not even live here.

A dive industry worker echoed these opinions:

The Pecelan Laut. They are the "face" but not the "substance." The idea of a "community-based" organization is not real. At the end of the day, resort owners are in charge. Look at the people who bought hotels here in the 1980s. That is who is really in charge. They have the money, they have to bribe local officials for us, and they bribe the head of the village too. Conditions on the reef are good and tourists come. Their effort isn't bad. Some of these owners are really good people, It's good for us, good for the environment. But it is not what I would call "community-based."

Several critics of the MPA and the Pecelan Laut said that management activities were really conducted to gain attention from international aid organizations offering awards and grants. Many of these respondents cited the weekly boating patrols that took every Saturday and Sunday to ensure the safety of the reef as an example. While a majority of interview informants felt this was a positive example of management work, several respondents with high community standing disagreed. For example, one stakeholder remarked,

"If the bad guys know you are coming on patrol, they will not be poaching on Saturday and Sunday. They will poach on Monday." Others echoed a similar sentiment, suggesting that the real purpose of the Pecelan Laut was to "drink coffee and smoke cigarettes on the beach."

Several TEs contested the idea that the local MPA manages the reef, often stating that they (TEs) are actually the ones who tend to impact reef conservation decisions through bribes and payoffs to local decision makers. TEs frequently noted that this is normal in Indonesian governance and has been occurring for decades, before and after the Reformasi. If a dive shop, for example, wanted to start a conservation program, install a coral nursery or bring partner NGOs into the village to do so, bribes would have to be paid to the village government and the Balinese provincial government in order to do so. One respondent noted, "Eco-conscious resort owners cannot live there forever," casting doubt on the long-term feasibility of this regime. Several respondents informed of the research questions for this project even argued that TEs implement a top-down pattern or reef management. Many respondents partially disagreed saying that although this may be the case now, increasingly real power has shifted and will continue to shift to the Pecelan Laut. "Give them a break," said one respondent who had owned a dive shop for many years. "Many of them are local guys and have never been to school. These things take time."

Some respondents felt that questions on whether the Pecelan held real power missed the bigger point. Their presence had resulted in real changes in local and regional mindset, especially regarding destructive and illegal fishing. The community has, in the past decade, seen the real-world economic potential of an eco-tourism approach, including how visitors are willing to pay more to stay in a community with strong stewardship credentials over its reefs. This perception among tourists increases with each international conservation award that locals receive. Many stakeholders argued that reef conservation and the Pecelan Laut are perceived as synonymous with local wealth generation in the minds of many villagers. Thus, even if TEs retire and move away, some say the regime will continue as is and that fears of the fragility or reef protection in Pemuteran are exaggerated.

I want to emphasize that although some important interviewees said that the Pecelan Laut do not manage reefs, the Pemuteran Pecelan Laut have widespread admiration in their community. The Pecelan Laut are a source of pride for many stakeholders. Its members tend to be highly respected, relatively prosperous young men who "had a place for the reef in their hearts" as one member put it. Members typically congregate in traditional dress at the beachfront office, where they encourage tourists to visit, make donations, ask questions and buy souvenirs. These villagers are there to make the socioeconomic links between village life and reef health apparent to visitors, even when real management authority and capacity may be limited.

Pemuteran's healthy reefs and its global reputation for sustainable development would suggest that the Pecelan Laut's self-perceptions as autonomous MPA leaders in reef management may matter more to the village than actual autonomy over decision making around the reef. These self-perceptions combined with their elite status in the community are possibly more significant than veritable decision-making autonomy in management. It is important to note that positive perception of the role of the Pecelan Laut was highest among community members who were native to the village, meaning that the Pecelan are a culturally acceptable form of leadership, which garners palpable legitimacy and, as such, has created social norms in the village around reef management.

6.3.5 Indonesia: Strength in mandatory membership

In other Indonesian field sites, members felt that mandatory membership was part and parcel with the MPA's legitimacy. Lovina's reef management organizations, by contrast to the Pecelan Laut, have mandatory membership for anyone operating a boat for reef tourism purposes. Respondents frequently mentioned words such as *strength* and *trust* when describing how mandatory membership enhanced legitimacy. "We wouldn't believe in the organization if everyone did not have to join," said one respondent.

As part of membership, all members are required to pay monthly dues in the form of a percentage of earnings and readily acknowledge that this is to avoid free riders and enhance "positive feelings" about the MPA organization. Stakeholders use dues for insurance and for infrastructure upgrades such as buoy replacements on local snorkel and dive spots. Members exhibited a large degree of familiarity, trust and friendship with one another. The payment of dues to the MPA institution was seen as a way to increase buy-in primarily through the emergency insurance program that would offer loans and grants for broken boats. In the words of one respondent, "I pay into the organization each month. I trust them, and we are the same religion. Working together makes us strong, and if we have a bad year we work together to get out of it."

6.3.6 Indonesia: Again, where's the science?

There was one major criticism regarding the legitimacy of Indonesian MPAs: a lack of science. The scientific landscape described by stakeholders within and outside the MPA was a patchwork quilt of NGOs, universities (both domestic and foreign) and some token but lackluster effort on the part of the Balinese governments to monitor ecological health of the reef systems on the islands. Stakeholders generally viewed the influx of well-known, international environmental NGOs as positive events. Many described conservation interventions in host communities such as citizen science-oriented monitoring as positive gains for local people. For instance, an Australian university partnered with the Balinese branch of several international NGOs to implement a community-based alert system regarding coral bleaching outbreaks, as well as a central database for tracking the bleaching and its intensity. In a dozen interviews with key informants who had participated in this program or programs such as this, there was often a fundamental gap. Stakeholders did not understand what to do with the data once they had collected it and handed it off to partner agencies. This type of feedback oriented learning and the uncertainty on the part of communities suggest that although legitimacy is high in co-managed systems, a fundamental weakness is the lack of a coordinated scientific entity linking science, policy and management.

In this section, I described how Malaysian MPAs are characterized by a lack of legitimacy among stakeholders mainly because they see the Marine Parks

Figure 6.8 An ecological training event in Amed held by several NGOs and a university partner. The intention was to train local dive industry workers to identify signs of bleaching.

officers as absent and bemoan the lack of conservation science. Stakeholders question the legitimacy of an MPA where its employees are paid high salaries, perform minimal enforcement duties and do not conduct scientific monitoring. Based on my interviews, there is a major need for research on the impacts of the dive industry on coral reefs in Malaysian MPAs. The dive industry is generally perceived as an environmentally friendly alternative to the fishing sector, but my observations saw many rogue dive shops allowing destructive behavior on reefs among their employees and their clients. Additionally, respondents confirmed my own observations citing a need for up to date science on coral damage in the new reef tourism economy.

Indonesian MPAs had a much higher degree of legitimacy among its respondents because the MPA was widely seen as the source of reef protection in these communities. Much like Malaysia, Indonesians also worry about the lack of science. Stakeholders feel that they have worked hard to form MPAs and stop destructive behavior on reefs but still need a systematic scientific program on the island of Bali. While NGOs are an important contributor, ideally the local-level government can and should provide this service given its authority, resources and its ability to overcome issues of fragmentation that trouble program-focused NGOs. The local government forming a scientific office, attracting the best marine science students from Udayana University in Bali and implementing a monitoring program on the reefs designed to inform local MPAs could provide longevity and continuity to the program-based scientific monitoring schemes that characterize NGOs. There has already been evidence for this type of reform in the 2011 Bali Rapid Marine Assessment report by Mustika, contracted by the local government, this detailed assessment of Balinese reef ecology is unprecedented in its breadth and depth. It is evidence that the local government can possibly contract with scientists to perform monitoring if setting up an office for scientific management were not possible.

6.4 Different Perceptions of Institutional Value

The second survey question on legitimacy asked respondents whether or not they value the MPA. There was a significant difference between Malaysian respondents (14 percent value their MPA) and Indonesian respondents (76 percent value their MPA). Interviews suggested two reasons why Malaysians did not value their MPA. These included the perception that Marine Parks officers were largely absent from their duty stations especially in the off-season, which led to illegal fishing, as well as the perception that the conditions of the reefs were rapidly degrading as years go by with little intervention from Marine Parks.

6.4.1 Malaysia: Hardly working?

Many stakeholders in Malaysia, when interviewed about whether they value the MPA, replied that they do not because many employees of the Marine parks were seldom working. One respondent, a well-known conservation leader in the community, related an instance where the Marine Parks headquarters in Kuala Lumpur learned about workplace absenteeism in Marine Parks. As a response, they installed biometric scanning devices at the offices:

> The result was that our guy [local Marine Parks Officer] would show up, scan his fingerprints, and leave. He always said he had to take his wife to work, but was he taking her to work for the whole day?

This view was widely shared among respondents, often sparking very heated exchanges from stakeholders when interviewed. Additionally, independent research by the UNDP confirms these findings. For example, in its mid-program report for its *Conserving Marine Biodiversity Through Enhanced Marine Park Management and Sustainable Island Development Program*, despite two years of program implementation meant to improve enforcement in Malaysian MPAs, their research found enforcement is still lacking (UNDP 2011). In my own interviews, one of the most widely voiced concerns regarding workplace absenteeism related to the seasonality of the tourist industry in Malaysian Marine Parks. Respondents spoke on the prominence of workplace absenteeism during the monsoon months. During the monsoon, many dive shops close or undergo regular maintenance on facilities and equipment in preparation for the next tourism season. Since many of the dive shop owners keep a close eye on illegal activities on the reef, informants spoke of Marine Parks officers completely deserting their posts. In addition, many informants spoke of illegal fishing within the Marine Parks boundaries occurring around the clock, during each day of the monsoon season. One stakeholder, who had lived in the nearby village for decades, said, "As soon as the last tourists leave the islands, in come the fishing boats, and Marine Parks officers, if they are here, only talk to these fishermen if they want a bribe." Informants repeatedly pointed out that these fishermen were not even locals from the villages within the Marine Parks themselves, but outsiders from nearby villages or from abroad. In the words of one respondent, "These reefs are not theirs, but we get to deal with their fishing lines and nets every year when we come back after monsoon season." This same respondent noted that he and his dive company once hauled a nearly 5,000 pound net off of one of the most popular reefs right before the beginning of tourist season.

Many did not value the MPA because they perceived reef health as rapidly declining. Stakeholders who had lived in MPAs for more than 10 years addressed the deterioration of the reefs, blaming both man-made and natural causes. They felt that the man-made causes of degradation were preventable and were the responsibility of the Marine Parks. Stakeholders commonly cited examples such as repeated anchor damage from smaller vessels on reefs high in tourist demand. One dive shop owner recalled leading a group of divers across a reef only to narrowly miss having an anchor dropped directly on her. "I watched it sink to the bottom after nearly taking my head off, and smash into a 2 meter stand of *acropora* corals. Where are the Marine Parks guys to stop this?" Her junior colleague added, "I have only been here five years, but I have noticed a sharp decline in the corals since I arrived here." His boss then replied:

And I have been here for fifteen years and can tell you that it is worse now than it ever was. Corals are being trampled, smashed with anchors, damaged by Go-Pro users with selfie sticks and Marine Parks is never on patrol to catch it. It is worse than I have ever seen it.

I then asked her why she thought that man-made damages were on the rise. She cited the widespread use of digital cameras in the parks, mixed with a lack of enforcement:

It's because on one hand you have more people coming now than ever, and you have dive operators that allow tourists to do whatever they want on the reef. Often the fact that everyone has their own camera is a huge problem, people touching, breaking, and pulverizing corals in order to get the perfect shot. Marine Parks should be educating people and citing dive operators who allow this. But where are they?

Several respondents from Marine Parks pushed back on this idea. "We cannot be at every dive site every day. Dive shop owners and employees need to brief their customers on the proper behavior on the reef." Another Marine Parks staff member noted that a major push of the Marine Parks was to hand out materials to local dive shops that they can post or distribute. These materials instruct divers on how to behave on reefs. "We do need to update them and bring attention to the problem with [digital cameras] and the more widespread use of cameras, but we think we do a lot of work on this issue."

6.4.2 Indonesia: Reefs as income generators

Stakeholders at the Indonesian sites valued their reef management institutions. Three key themes emerged to explain this. The first was that many considered the reef, the village's well-being and the institution as a fundamentally linked system. Second, cultural and religious ties to the reef management institutions were a source of value for respondents. Last, the idea of self-sufficiency in a nation haunted by a history of corruption was a widely held source of value for stakeholder respondents.

Stakeholders when asked if they value their reef management system in Indonesia said yes by a wide majority. The first reason for this perception was a belief that the reef and local ways of life were intrinsically linked. "We are the reef and the reef is us. We are in charge of these reefs, and the village takes the responsibility seriously because this is our income" was the view of a longtime dive resort owner in Amed, Bali. This view, where the reef ecosystem and the economy are very closely linked, thereby inspiring widespread buy-in for conservation, was nearly universal among respondents. At times, interviewees stumbled on the question Who manages the reefs in this village?" because it seems so obvious to them that it is the village that they could not see why an interviewer would even think to ask this question. One respondent's response, "Who else would or could manage our reefs?" best sums up this perception.

Stakeholders who had lived and worked in reef tourism for many years tended to point out a pattern whereby enforcement for poaching or destructive fishing practices was needed much more in the 1970s and 1980s than it is today. One stakeholder told the following story to illustrate the difference:

> In the late 1980s when I bought this place, dynamite fishermen would come over in boats from other villages around Bali or from Java, or even as far away as Thailand. Anyways, I bought the biggest shotgun I could buy on the black market, and would just go out in my own boat, point the gun at them, and tell them to leave. One time, they came back with torches and wanted to burn my hotel down, but the local village leadership intervened and it didn't happen. Enforcement back then was completely lawless, but nowadays, there's no need for this level of conflict. People get it now, mostly.

This conflict is no longer the way of life across the Balinese villages in this study, due to increasing awareness. The most important factor behind this awareness according to many stakeholders was the ever-increasing links between rising incomes and reef tourism. In the late 1980s, with the spread of this type of awareness, traditional associations of fishermen, known as *kelompok nelayan,* began to assume the duties of reef management as they ran dive or snorkel tours in between fishing trips. Ultimately, many were able to learn enough English to abandon fishing altogether and instead lead tours, ranging from diving and snorkeling to sunset tours to tuna fishing tours. In the words of one fisherman-turned-snorkel tour boat operator in Lovina:

> I used to make no more than [$5] a day fishing. I now can make [$50] in a day if I can get enough tourists on my trips. The choice is clear. You can get lost at sea, you can go out for days and not catch anything, I stopped fishing nearly 20 years ago for obvious reasons. My life is a lot easier now. I am not rich, and I do fish when tourist season is really slow, but it is not the majority of my income anymore, and it is the same with the other guys I used to fish with. The only people still fishing are very old.

6.5 Sharing Power with Stakeholders

Regarding power sharing, I asked Indonesian stakeholders a different question from Malaysian stakeholders. Power sharing is an important component of legitimate institutions, but in Indonesia and Malaysia it looks very different. In Indonesia, power sharing devolves decision-making authority to villages, while in Malaysia it would simply involve information sharing from MPA officers to stakeholders. I asked Malaysian stakeholders whether the MPA distributed information to them, and 76% of respondents said no, compared to Indonesia, where 89% of respondents said there was a form of power sharing between government and the village.

6.5.1 Malaysia: When information is not enough

In Malaysian MPAs where 76 percent of respondents said there is no information distributed, this was surprising because it directly contradicted my observations. In every restaurant, hotel and place of business across the Perhentian Islands and Tioman Island, informational posters distributed by Marine Parks officers are on every wall (Figure 6.9). Why then did stakeholders say that Marine Parks did not share information with them?

One does not need to look very far in order to see a leaflet, poster or sign from Malaysian Marine Parks. According to Marine Parks officers, they are required to be posted in every hotel, boat office and business on the islands. They are written in every language of potential visitors: English, Malaysian, French, Russian, Chinese, Japanese and Korean, and they have illustrations of different species of coral and fish, as well as proper behaviors for snorkelers

Figure 6.9 A poster with rules and regulations of Malaysian Marine Parks in the state of Pahang, Tioman Island. It informs visitors that there are rules according to Malaysian law that must be followed in the park, what the key reef sites are and what the prices of admission are.

and divers. These include not touching or standing on corals, not harassing the wildlife in any way and not leaving any rubbish on the beach or the reef itself. That said, stakeholders nearly unanimously said that the reason they said that no information is given to them is that the information campaign has had little observable effect on the behaviors of visitors.

Commonly mentioned behaviors include touching corals, trampling corals, hitting corals with dive tanks in order to take a photo and those who cannot swim yet hold a recreational diving license (a common phenomenon in Southeast Asian recreational diving community) holding onto corals in order to propel themselves forward. At least one of these behaviors was witnessed on every survey dive that occurred for this research. Dive shop owners would frequently mention the tension between doing business and preserving the reef for future tourists, and the limited ability of the information sharing to help resolve it:

I have seen things on these reefs that you wouldn't believe. Just this morning a whole group was walking across a massive field of acropora corals. This group was 12 people to one dive master. That's too many people. Nobody follows the signs and posters in this park! I know it doesn't make sense to give a dive license to somebody that cannot swim, and at my shop we don't. But for every 1 person we turn away who can't swim and would spend their whole session drowning and clinging to the reef with a diving regulator in their mouth, the dive shop next door will certify 10 of these types. So if you go into this with the assumption that this person is going to get their license regardless it doesn't make much sense to turn anybody away. Add in the fact that recreational agencies like PADI require us to give licenses to so many people a year and you have a recipe for disaster. The incentive structure is all wrong, for both divers, dive instructors, and the whole industry and the reef bears the burden.

When I asked this respondent to describe the problem in greater detail, they explained that there is a new influx of tourists in Southeast Asia thanks to the rise of the East Asian middle class, with greater disposable income. The same respondent said the following, which was a remarkably representative and widely held perception of dive industry stakeholders as a whole, including those from Indonesian villages surveyed for this study:

We are seeing a massive influx of people from East Asia, specifically China and Korea, as we've always had Japanese tourists. These cultures do not have the same education opportunities around wildlife viewing that tourists from Europe and America have had. Most, that's not to say all, but most East Asian tourists consider touching the wildlife and taking something to remember the reef by as an added bonus to the tour. Many of them will seek guides and shops that condone this behavior. On the island, we know who they are, we know the guides that do this, and they make a lot of money doing things like letting people keep starfish. A poster is not going to stop this behavior, it is culture.

A different stakeholder respondent echoed these same perceptions, while describing the conflict between information posted and the motivation behind acquiring a dive license for many tourists:

A big issue for the reefs is the Chinese and Korean tourists who seek recreational licenses without knowing how to swim. Because swimming is not the same type of culturally widespread recreational activity as it is in Western countries like where you're from. That said, diving is a recreational activity that shows prestige and wealth. Certain dive shops in certain parks get reputation for certifying literally anybody and these shops are very popular among large groups of tourists from Mainland China in particular. Those tourists will not look at a poster and say, "okay I will respect nature now."

When I asked stakeholders about the efficiency of the informational material, the nearly unanimous perception was that there was no noticeable effect. Several divemasters and dive shop owners interviewed noted that they lived and worked on the islands long before these posters appeared, and they had little to no effect. One divemaster had the following to say, "If you think a poster, even a poster written in Chinese, is going to change mindset in the course of the two minutes it takes people to read it, then I don't really think you get people." Likewise, stakeholders consistently pointed out the mismatch in the mindsets between those who visit the reef on a family holiday, as most the visitors and divers do in Malaysian Marine Parks, and the scientific outreach and educational components contained within the posters. One prominent dive shop owner's comments summarized the mismatch, "Nobody comes on holiday to learn a biology lesson. And even if they do, nobody is going to miss an amazing photo to protect the coral. People still do not realize the sensitivity of coral, they think of them as rocks."

It is important to point out that some stakeholders disagreed with the perception that the Malaysian Marine Parks materials had little effect. One Marine Parks officer discussed an example where he heard first-hand that the posters had informed someone for the first time that corals were not rocks. I asked this Marine Parks officer why there was such a widely held perception that the informational material did not have an effect, and he also blamed the mismatch between being on holiday and behavior change, saying, "Marine Parks tries, but ultimately it is up to the individual person to follow our rules. We can't monitor every dive, this is also a responsibility for dive masters and dive instructors. They need to educate and enforce good behaviors." The comments of this Marine Parks officer hinted at an interesting conundrum, which I did witness dozens of times on recreational dives in the Malaysian Marine Parks. Even the stakeholders who had the most seasoned records of conservation and outreach would be seen to, at times, look the other way and allow their guests to touch the corals. The importance of the Southeast Asian cultural norm of "saving face" is a plausible explanation for this sort of behavior, where it is considered poor manners to contradict somebody else's behavior in a negative way. That is not to say that this type of incident happened on every dive, and when it did occur in very extreme ways, for example, somebody lying down on a bed of corals, it would provoke a rebuke from the divemaster. However, using plating corals as a way to control momentum in the water was an all too common observation, which rarely drew reaction from dive leaders.

When I asked stakeholders what they thought may be a more effective educational tool, many stakeholders were at a loss. There was a degree of consensus that Marine Parks officers should do "underwater monitoring." Given that these parks are small and there are less than a dozen dive operators on each of the islands, people tend to know one another. Thus, it was suggested that if there was some enforcement capacity among Marine Parks to observe divemasters who frequently allow their clients to break the rules and touch the corals, they can be issued tickets. In the words of a proponent local leader on these types of measures, "Right now dive operators see allowing this behavior as a good business strategy. We need to find a way to make it a bad business strategy where the dive industry helps in enforcement." However, those who did suggest underwater monitoring acknowledged that it was unlikely because of the capacity gap of Marine Parks. One stakeholder illustrated the perception in the following statement, "How can we expect guys who never come to work to patrol underwater and issue citations? It would never happen. And even if it did, you bribe them with 20 ringgit [5 dollars] and it goes away. This is Malaysia!"

The second theme in interviews was the historical lack of community consultation in the early years of Marine Park formation that many felt was far more important than posters and information. This finding is supported by those of Islam et al. (2014) and UNDP (2011), studies that also saw a lack of historical community consultation as a major barrier to effective management in Malaysian MPAs. Take, for example, the formation of the Tioman Island

Marine Parks, where a handful of historical fishing villages became a marine park "overnight," according to many respondents. One stakeholder spoke of the overnight transition, "One day in the 1980s, government officials came to the island and told us we weren't allowed to fish the reefs anymore. Families here had fished for generations and then overnight it was banned." These statements are supported by Abdul (1999), who states that Malaysian fishermen settled Tioman Island several hundred years ago, with the federal government gazetting Tioman Island as an MPA in the mid-1980s.

Respondents noted that through time, the initial resentment felt by the community toward the Marine Parks died out: "Most of the original generation of fishermen have retired or passed on, so few people remember the anger felt by the community." Nonetheless, problems still linger. The Malaysian government has, according to interviews, sought multi-lateral aid in order to build capacity for community consultation. This 2007 program, though the UNDP, and called the *Conserving Biodiversity through Enhanced Marine Parks Management and Inclusive Island Development*, seeks to build community consultation capacity in the Marine Parks Department in the age of rapidly expanding tourism. The majority of stakeholders who mentioned the program described it as well financed, well intentioned, yet poorly executed. The words of one conservation expert are representative of many Tioman stakeholders:

The UNDP project was set up in order to address the fact that government came in a few decades ago and told fishermen they can no longer fish, thus ending generations of livelihoods and traditions. They trained a handful of fishermen as boat guides, but most people around here who have sought to convert from fishing to tourism have done so a generation ago. It is outmoded, better suited for this community 25 years ago.

Another informant who spent years working in the dive industry and is now retired and working in conservation added the following,

The UNDP Program sounds great, it has all the right buzz words right? Livelihood transitions, capacity building, community consultation, education programs for locals but the fact of the matter is other NGOs have been doing this for ten plus years in the area, and often, the Marine Parks Department impedes any real progress. Since they were the main recipient of financing and training through this program, you'd think you would see a change in their absenteeism, or their lack of enforcement for illegal fishing, but you didn't. The worst part was, they featured some of the most flagrant work dodgers in the news write ups about the success of the program!

Many informants shared the views that despite this high profile and publicized program, it had limited effectiveness. Many said that the main managers in Malaysian MPAs were the handful of people working for Reef Check Malaysia. This study found that nearly everyone interviewed, when asked who monitored the reefs, replied in some way that Reef Check Malaysia was performing the real day-to-day conservation work as well as the community engagement work. Several stakeholders commented on the sensitivity of this widely held perception, "Marine Parks knows people think they do not do their job, they also know we think Reef Check is the major conservation player here, so they try and limit what Reef Check can do. Last year they denied them a desk in the main office. Petty stuff like that." One stakeholder related the story of a massive fish die-off, saying that whenever illegal fishing boats anchor offshore just outside the Marine Park borders, they often discharge pollutants resulting in a fish die-off the following day. This stakeholder remarked quite confidently, "We don't call Marine Parks when this happens. You know who we call? Reef Check Malaysia, because we know they'll answer the phone and they'll at least try and so something about it." When I asked Reef Check Malaysia what it is they can do, they replied with disappointed hesitation, "We try as hard as we can to relay this information to decision-makers in Kuala Lumpur. But where it goes after that, it isn't clear. This is why our community engagement programs are so important, because of the slow speed of the bureaucracy."

The final theme that arose regarding legitimacy in Malaysian Marine Parks was a near consensus that although the Marine Parks were largely failing in its management and conservation aims, most stakeholders believe that it is government's responsibility to manage the reefs. Stakeholders ranging from boatmen to long-time village leaders held the perception that in Malaysia, the government is in charge of its natural resources and any other arrangement would lend the system more corruption and inefficiently. In the words of one respondent who had worked on the Perhentian Islands his whole life, "Malaysians are like Americans. We view government as the one in charge. The villagers lost any control years ago when parks were formed, placing us back in power wouldn't make sense." The literature on Southeast Asian marine governance supports these findings citing political events such as the cancellation of local elections in the 1970s, as well as laws and regulations such as the National Security Act, which limit popular participation as reasons (Pomeroy 1995). Additionally, freedom of assembly is not widely considered to be a right and is limited by government. Furthermore, studies have found the political and regulatory framework in Malaysia could not support community-based management and that former fishing communities, such as those present on the islands in this study, lack the organizational capacity to build necessary institutions for localized management (Pomeroy 1995).

What does appeal to Malaysian stakeholders, as seen in this study and others, is a level of comfort regarding the sharing of certain responsibilities, primarily on enforcement of rules. In the words of several stakeholders, one possibility could be reporting dive outfits that repeatedly allow their clients to damage the reefs to the Marine Parks Department. Another could be offering compensation to the dive shops when they are asked to perform management and maintenance work for marine parks. Another suggestion included requiring all dive shops to give clients a statement to write out and sign before they dive where they agreed to not touch, stand on or disturb corals, lest they be cited by the Marine Parks. Respondents suggested that having people see and write out the rules might help them remember it under water and not have an adverse reaction to divemasters correcting bad behavior under water. Additionally, if they perceive the possibility of sanctions for rule breaking, they may try not to do so.

Regardless of how the shared enforcement would look, and despite the repeated commentary on the failures of Marine Parks, there is a persistent perception that greater sharing of responsibility is needed. Several stakeholders with marine science and conservation backgrounds pushed back on this however, citing past examples of programs where they, through their NGO or educational group, sought a permit from Marine Parks to carry out an ecological survey. They subsequently found their request for a permit denied. The reasoning, they argued, was that Marine Parks did not want to appear as though they were outsourcing their responsibilities to third parties. The official reason handed to the applicants was that Marine Parks felt they could not maintain their project. One stakeholder summed up his organization's struggle, "We all have degrees in biology, and were denied a surveying permit, because they said we had no training. I just am at a loss for words about it."

6.5.2 Indonesia: Genuine leaders

Among Indonesian stakeholders by contrast, 89 percent felt that there were genuine power-sharing mechanisms in place for them to manage local reefs. Stakeholder perceptions regarding power-sharing were generally positive with a few important criticisms that warrant discussion. Two themes characterized Indonesian stakeholders' views on participation. The first was a high degree of importance placed on hierarchal structure and leadership of the MPA. The second was a sense that despite efforts aimed at conservation, Suharto-era corrupt networks exist.

Indonesian stakeholders perceived that management authority was held by their village leaders, and this therefore lent them legitimacy. A very important theme to Indonesian respondents was that of leadership and the legitimacy of its conservation leaders as role models for conservation. Similar studies have demonstrated a link between healthy ecological outcomes, co-managed institutions, a well-defined leadership within the institution and legitimacy (Pomeroy 1995). Many respondents would list for me the details of the MPA hierarchy. They would describe in detail who held what job, how that person could acquire the job, what type of background they had to have and what responsibilities they had. The hierarchy and the structure of the MPA is widely regarded as important to the community, and this was reflected in knowledge among stakeholders of who was in charge of doing what.

Lovina, for example, had stakeholders who were able to name a single MPA leader and the leaders of MPA organizations in the other villages. Furthermore, every member of the organization is expected and encouraged to lead the organization at one point in his career. The idea of what constitutes a member in an Indonesian MPA varied from site to site. In Lovina, all men who worked in tourism were required to be members in the MPA. This requirement meant that they had to pay dues and attend monthly decision-making meetings. Membership in Pemuteran was voluntary, and anybody who worked in reef tourism could join. Membership in Amed was similar to Lovina's whereby anybody who owned a traditional boat, or a jukung, had a space in the fishermen's cooperative or kelompok nelayan and therefore could vote on issues in the MPA. Heads of reef management institutions in four of Lovina's six villages unanimously touted this opportunity for leadership as a source of legitimacy among members, saying that ultimately serving as leader means you have all the more reason to respect the current leadership. Respondents argued that transparency, trust and pervasive feelings of ownership among stakeholders were linked to the possibility that they too could be the leader one day.

Several respondents noted that despite local leadership, corruption was still an issue and local conservation leaders were often those who were willing to pay bribes to local leaders to get things done. "It is the TEs who give the money for real monitoring, enforcement, and the necessary bribes that ensure smart planning for reef conservation." In the words of another respondent, "You can look at leadership in two ways here, formal leadership of enforcement on the reef, and informal leadership of bribing people who stand in the way of conservation and need a pay off." Many other respondents echoed the belief that outside of the formal MPA there was a network of wealthier elites in the community who took on the cause of conservation and would pay off people in order to make conservation planning happen.

6.6 Summary and Conclusions

This chapter described how stakeholders perceive their MPAs to be more legitimate when they are co-managed. In the beginning of this chapter,

I introduced a six-part definition of legitimacy that included (1) whether the MPA manages the reefs, (2) whether stakeholders value the MPA, (3) whether power-sharing arrangements are in place, (4) the use of science in decision making, (5) the visibility of MPA leaders and their frequent interactions with stakeholders and (6) leaders and members of the MPA taking responsibility for their role as reef protectors in the village.

Starting with Indonesia, stakeholders believed the MPA manages the reefs, they value the MPA, they perceive proper power-sharing arrangements are in place, they appreciate MPA leadership because of their presence in day-to-day life in the village and they appreciate the responsibility that local MPA leaders and members take for the reef. Science behind rule making is a major point for improvement as is local level graft. Malaysian stakeholders on the other hand do not see the MPA as managing the reef, do not value the MPA, do not see the current form of power sharing as useful or effective, do not see local MPA employees interacting with stakeholders very often in positive ways and do not see MPA employees, officers or leaders as taking responsibility for the reef. Like Indonesia, Malaysian stakeholders also lament the lack of systematic scientific study and monitoring on their reefs. Malaysian Marine Parks must find real ways to address absenteeism. This could include reforming the way Marine Parks hires employees, changing it from a single civil service process to its own process meant to attract people with marine science training. In addition, Marine Parks officers should engage local stakeholders in order to share some management responsibilities in a more formal way. Examples could include reef cleanups or issuing permits for ecological data collection.

In Section 2.3, I discussed how multilateral aid agencies have been shifting resources to capacity-building programs focused on co-management despite the criticisms that there is a lack of studies that demonstrate whether hypothesized benefits (such as legitimacy and empowerment) were observable (Cinner et al. 2012; Wamukota et al. 2012; Stevenson and Tissot 2014). This chapter shows that co-management does very well in five of six components of legitimacy. Legitimacy is also important because it informs whether local people follow rules and carry out difficult sanctions or alternatively break rules, illegally fish and dodge their sanctions.

Chapter 7

ADAPTIVE CAPACITY OF MARINE PROTECTED AREAS

7.1 Overview

The previous chapter discussed the significant differences in perceptions on legitimacy. Whereas stakeholders in co-managed settings attributed high degrees of legitimacy to institutions, the opposite was true for stakeholders in top-down marine parks. This chapter discusses the final set of differences in stakeholder perceptions: adaptive capacity. Stakeholders in co-managed settings felt that they were better able to learn and change management strategies when the need arose. In this section, a major component of the definition of adaptive capacity is organizational learning. Organizational learning is how a marine protected area (MPA) acquires understanding, know-how, techniques and practices. The actions of individuals are very important in a dialectical sense, because their own learning feeds the organization's learning just like the organization itself impacts the ability of individuals within it to learn. The result of learning in adaptive capacity must be innovation, or the crafting of improvements based on learning.

Much like in the previous chapters, I will present two hypothetical cases, Case A where there is an MPA with adaptive capacity and Case B where there is not. Beginning with Case A, stakeholders in this MPA frequently talk about man-made environmental degradation that troubles their local reefs, but they also talk about examples of how these mistakes were fixed in the past and in the present. For example, stakeholders discuss rule breakers who fish for snappers and groupers under the cover of darkness and strip these important ecological functional groups off the reef. Once it became clear, through people reporting this behavior to the MPA leaders and members, action was taken to increase surprise patrols on the reef at night. Violators were caught and forced to pay fines. If they were caught again, they were reported to the police. The village is small, so night patrols could be funded through extra fees paid by hotels and dive shops. In addition, the small size of the village makes it so that offenders are largely known and personally asked to stop rule breaking by friends and neighbors working for the MPA. For trickier environmental problems such as annual bleaching or crown-of-thorns starfish outbreaks, innovative local actions are developed that may not reverse the trends but play a role in mitigating the damage and raising community pride in taking action. For example, in years of extreme bleaching, the MPA in Case A may limit visitors to thermally stressed reefs. In months with intense outbreaks of crown-of-thorn starfish, local men may use spears and spend the weekend culling them. Stakeholders in Case A mention how, through the years, they've changed certain programs that did not work. For example, when too many people were seen to be crowding the reef, they enacted a series of limitations and an ordered schedule of access for locals to access the reefs with tourists. Finally, stakeholders describe the wide range of possibilities for creative innovations that combine tourism, conservation and local culture. For example, many local businesses fund reef cleanups linked to local religious holidays and cultural festivals that attract environmentally conscientious tourists.

Case B is an MPA where there is minimal adaptive capacity. Stakeholders frequently talk about man-made environmental degradation that happens year after year that they cannot fix. For example, it is known that large vessels frequently break rules and fish the MPA, but nobody in the MPA stops them. Stakeholders also mention annual ecological crises such as bleaching but are unable to close certain areas for diving and snorkeling even in periods of peak bleaching and stress. Programs that do not work are left in place and not updated or changed. For example, there are signs everywhere asking people not to litter the reef, yet littering continues to be a large problem in the MPA. There has been no discussion or reflection of the problem, or how the MPA managers can supplement signs with additional efforts aimed to stop littering. Finally, many local business owners would like to offer programs or attractions such as coral nurseries or artificial reefs, partially in an effort to balance the recent uptick in coral cover loss, but these ideas are squashed by the MPA, which is unable or unwilling to permit or assist in the implementation of these ideas. Case B touches on some additional topics of this chapter: obstacles to adaptive capacity, including corruption, confused bureaucratic responsibilities and inadequate training of those who work for the MPA. The following sections display the survey results and discuss findings on adaptive capacity in greater detail, and describe how the actual case sites may resemble these models.

7.2 Survey Results

Figure 7.1 shows the statistical tests that I completed on three questions meant to query stakeholders on adaptive capacity.

Something bad happens on the reef (such as anchor damage to corals that requires the MPA to act). Do members of the institution reflect on past management interventions before deciding on new ones?

	No	Yes	Total
Malaysia	83 (0.99)	1 (0.01)	84
Indonesia	37 (0.26)	108 (0.74)	145

$$p = 0.000***, z = -10.70$$

The reef management institution sees that one of its programs is not working. For example, it learns that large amounts of poaching happen during a certain time of year. Can it change its program to address this problem?

	No	Yes	Total
Malaysia	73 (0.78)	21 (0.22)	94
Indonesia	33 (0.22)	114 (0.77)	147

$$p = 0.000***, z = -8.42$$

Does the reef management institution support businesses or stakeholders who want to intervene for conservation in innovative or creative ways?

	No	Yes	Total
Malaysia	82 (0.90)	8 (0.09)	91
Indonesia	25 (0.20)	117 (0.95)	123

$$p = 0.000***, z = -10.84$$

Figure 7.1 Survey responses to questions on adaptive capacity. *** : $p < 0.001$.

7.3 Different Stakeholder Perceptions on Learning

There were significant differences in stakeholder ability to use previous knowledge in order to address current crises. Seventy-four percent of stakeholders in co-managed MPAs said that past crises were discussed before decisions were made about how to deal with current crises. Only 1 percent of stakeholders in Malaysian Marine Parks perceived that learning occurred in this way.

7.3.1 Malaysia: Fear and learning

Respondents suggested that there were two main reasons for the lack of institutional learning on past interventions in Malaysian Marine Parks: unpredictable sanctions or rules imposed by Marine Parks and confused responsibility that permeated the bureaucracy.

Respondents spoke often of crises such as the predictable resurgence of illegal fishing and poaching on the reef during the monsoon months when there are no tourists in peninsular Malaysian Marine Parks. "This happens every year, we know that a fleet of trawlers will come as soon as the tourist boats stop coming," said one retired conservation worker and former Marine Parks officer. Even though stakeholders across economic sectors, ranging from tourism to diving, in addition to Marine Parks officers, knew this would happen, nearly every stakeholder interviewed said that it was impossible to stop. Informants in tourism and the dive industry blamed a lack of predictable permitting or approval processes for proposals they have submitted to combat crises. Several nongovernmental organizations (NGOs) described applications that they had submitted specifically for projects aimed to address ecological crises based on knowledge of past crises that were rejected.

One such application was for a coral growing area meant to address the fact that many healthy reefs were repeatedly trampled on near shore areas. Malaysian Marine Parks rejected this proposal because the scientists who applied for the permit, who brought their own financing and long-term maintenance plan, were said to be "untrained" and "unable to maintain the nurseries." Another proposal mentioned by stakeholders was to implement extra litter removal programs after major holidays in the villages as a response to large amounts of trash appearing each year on the reefs. Several interviewees mentioned that Marine Parks officers, for reasons having to do with bureaucratic procedure, also rejected these programs. In the words of one respondent, "We just went ahead and did them anyways, they don't have the capacity or the work ethic over at Marine Parks to stop us from running rubbish removal programs in the village." Another informant, discussing a similar program that was rejected, said the following:

We definitely have stakeholders who have lived and worked here a long time, who can look back in time and say, You know, in El Niños in the past, it helped to close the really shallow areas in the park, because these corals were in warmer water to begin with, and could benefit from less visitor pressure on them, fewer divers kicking the corals and kicking up sediment. But when you bring this knowledge to Marine Parks, they send you around and around, to person after person. Nobody wants to talk to you, nobody knows who is in charge. Nobody can review yours and others' plans for action and say, "yes this is a good one!"

Respondents from within Marine Parks agreed that it was difficult to approve stakeholder requests that used events in the past to adjust current and future management. They did not see this as incompetence, however, they saw it as a confused and bureaucratic system where people with the ability to approve such programs work hours away from the islands in offices in Kuala Lumpur. Many Marine Parks officers described multiple attempts to gain approval for programs either within Marine Parks themselves or in partnerships with outside groups. One officer said the following, a sentiment echoed among others working in Marine Parks:

> A lot of locals who work here complain about us, and say we don't do anything. You're not the first person to write about Malaysian Marine Parks. I have seen other researchers that describe Marine Parks as ineffective. But I must emphasize the extremely difficult nature of getting approval for programs from central offices in Kuala Lumpur. People there are hard to reach and even harder to secure commitments from.

Other informants from within Malaysian Marine Parks noted that getting approval for even common sense programs at no cost to the Marine Parks is very difficult. Most commonly, this is because if the project were to fail, or were to offend the wrong people, everyone in that chain of approval could lose their job. Informants discussed pitfalls of programs as falling into two categories: offending powerful people or not succeeding and, as a result, getting the person who approved it fired. For example, when stakeholders mentioned the predictable off-season poaching that occurs in Malaysian Marine Parks, they noted that many of the owners of these fleets of fishing boats were extremely wealthy Malaysians. One stakeholder even said, "I have spent decades reporting this. In Malaysia, if you have money, any rule can be broken, and these fishing companies have a lot of money."

Another example of an obstacle to learning is the inability of stakeholders to prevent irrigation of golf courses on Tioman Island across several large resorts. Many stakeholders from both Marine Parks and the villages were aware that Tioman Island was in a severe drought and that water supply was at risk. Several took steps to intervene, through Marine Parks, through the local government of Pahang and through contacting the resorts themselves. The irrigation continued however, even for golf courses that would go largely unused in that particularly slow tourist season. Stakeholders repeatedly said that because powerful people own hotels, and they are closely connected to the even more powerful Sultan of Pahang, that their actions would go on without question. Many respondents mentioned how the large amount of fertilizers used on the golf courses is part of a wider issue: poorly planned irrigation. The nutrients in the fertilizer enter the coastal waters as runoff and have detrimental impacts to the coral environments, which require low nutrients. "We know these practices will impact the coral and our livelihoods in the long run," said a dive industry stakeholder, "but nobody wants to help us lessen the irrigation stress because of how well connected resort owners are." To add to this issue, Marine Parks officers, when asked how the problems with golf course irrigation and runoff impacted their own duties to ensure conservation on the reefs, replied that anything that happens on land is not under the authority of the Marine Parks and that they could get in trouble if they were perceived as intervening in landward affairs that were not within their job description. "It is not really clear what agency would need to address the irrigation and runoff problems here," said one Marine Parks officer. "Even if we did know, in Malaysia it is very difficult to get two agencies to work together because we all have different priorities and work descriptions. We could lose our jobs."

This fear of job loss hindering learning permeated the interviews of people working in Marine Parks or in government. It was specially prevalent in cases where authorizing a project, even though it could seem like a common sense project, may come back to haunt them. For example, a Marine Parks officer discussed the coral nursery example described above as a particular risk for him to authorize as it could result in job loss. "If I were to authorize a foreigner to come in here, even though he is a scientist, to build this nursery and it gets abandoned, then I am to blame for the costs of cleanup and any problems it creates." He went on to describe how Marine Parks is very wary of authorizing any projects from foreigners or from outsiders because this may be seen as Marine Parks' inability to do its own job. "We are paid to be out here responsible for conservation. If we give approval to third parties, we need to be very careful about it because it may make us look lazy or like we are not able to do similar projects ourselves."

Other stakeholders outside of Marine Parks pushed back on this idea arguing that officers use the excuses of job loss as a way to not have to spend any time or effort on the job. "What it comes down to is this," said a long-time resident who has a background in marine conservation, "Marine Parks guys see partnerships and projects as time consuming sources of extra work that will require them to leave their air conditioned offices in their massive building. This is why they reject them." This stakeholder and many other interviewees went on to describe project after project implemented by Marine Parks left to "rot in the water." For example, shortly after rejecting the permit for an NGO to install a coral nursery, Marine Parks on the Perhentians installed one of their own. Several months later, according to a half dozen informants, the corals were all overgrown with algae and dead because there was a lack of capacity among the officers to clean the corals and perform regular maintenance on the project. "Additionally," says another conservation group worker, "they sited the nursery in an area heavily trafficked by boats, so it is no wonder that the water quality was poor and the corals were smothered."

7.3.2 Indonesia: Learning with pride

Interviews with Indonesian stakeholders in areas with co-managed reefs told a remarkably different story, characterized by two attributes. The first was a profound desire to learn from the past and readily admit previous errors in management and the second was an importance placed on easy and unencumbered abilities in co-managed institutions for interventions aimed at conservation. For example, boatmen in Lovina discussed how the village with the healthiest reef tract for snorkeling had, in particularly stormy years, lost much of its coral cover. Anturan, the closest village in Lovina to this reef, allowed all neighboring villages in Lovina to bring tourists there, as long as the management costs and responsibilities were shared. One boatman, who had previously served as head of the local boatman's cooperative had the following commentary:

In the early days when we first gained real responsibility over the reefs, some storm seasons were worse than others. At first we would not limit the use of the reefs after the storms, but later we learned that if we closed down some snorkel areas, reefs would come back faster. We worked with other local villages to have them comply and help enforce the closures.

Another added, "We have very many rules in our village for who can bring tourists to the reefs. When we were arguing with other villages about who is breaking closure rules, we knew we had to write new rules to prevent the fighting." A third stakeholder concluded the thought, "Without these rules for ourselves and the way we interact with our neighbors, we would be fighting a lot. Before we had these rules we fought a lot, and now fighting is minimal. This fits better with our religion [Hinduism]." Informants mentioned how two decades ago, the argument over responsibilities could result in multi-year feuds between the villages of Lovina. The rule-making and institution-building, however, had put an end to this problem.

In other co-managed sites, stakeholders frequently mentioned how important "easy" intervention is, where learning can be applied without obstacles in order to change behavior or rules. Examples of obstacles included arbitrary rules placed by government, the need to give bribes or the village heads making rules that tied the hands of people with management responsibility. For example, several stakeholders discussed the ever-increasing problem of bleaching and its links to their occupations in the dive industry. They described an example of learning-based management where several Australian universities had partnered with globally recognized NGOs to implement a biodiversity-monitoring program in Bali. This program included a multi-day training both in a classroom setting and on the reefs. Dive industry stakehold-ers, particularly divemasters, were trained in how to spot corals in varying degrees of bleaching-related stress and how to count different genera of corals. The result was to increase the capacity of locals to notice when ecological problems were occurring and to send the data to a central database held by the university and NGO partners. This monitoring program was geared towards improving stakeholder-driven ecological knowledge in Indonesian villages that are facing the major challenges of climate change. "Nobody stepped in to prevent us working with this NGO," said one stakeholder. "Government does not interfere with us or our programs."

Another example of the importance of applied learning is the Reef Gardeners organization in Pemuteran, founded by conservation leader Chris Brown. Based at a well-known dive resort, Reef Seen, Brown and local partners formed this organization in response to the periodic outbreaks of crown-of-thorns starfish (*Acanthaster planci*). Local men from Pemuteran were trained to recognize these organisms, which feed on living coral and can rapidly destroy a reef, and harvest them from the reef using spears. The Reef Gardeners office space on the beach became a monitoring point and educational post for tourists looking to learn more about the periodic outbreaks and how the community deals with them. A well-respected local conservation leader, when asked about the program's impacts, said the following:

> We know that the science says that *Acanthaster* outbreaks are too many and too big to eradicate by hand, but this is a social program just as much as it is an ecological one. We are training local young men that in order to be respected in their community, they need to have an idea of what is going on in their local reefs, and they need to know how to intervene if there is a problem. There was a period where we saw massive outbreaks here. Locals expressed an urge to respond, and Reef Gardeners gave them that way, but in doing so it also spread knowledge, learning,and awareness.

Another stakeholder said the following about the program, "We saw these crown of thorns starfish killing our reefs and we acted. Nobody stopped us, we did it ourselves."

Indonesian stakeholders prized the ease of conservation action in co-managed reef systems and also were very open to discussing their own learning-induced shifts of opinion regarding conservation. Stakeholders saw a link between the ease with which the institution can intervene and learning from past mistakes or crises. Stakeholders in co-managed sites frequently mentioned their individual learning processes and past mistakes, mentioning their own changes in beliefs on what behavior could and could not occur on reefs.

This blended with conversations on institutional learning, which, according to many respondents, was a result of regular meetings among those in charge of reef management, and individual reflections and discussions were shared with the group. One example is illegal spearfishing at night. "I used to be of the belief that if I broke rules here and there it wouldn't come back to hurt me," said one local boatman in Lovina. "I used to take Australians spearfishing after dark when people wouldn't know. After time, I saw the way it damaged the reefs and stopped doing it. I shared my mistakes with my organization at our meeting." As he shared this, others sharing coffee smiled and laughed about the punishment he received for violating the rules and patted him on the back for coming clean. He continued while others nodded in agreement, "These are our reefs, and we can't do that to them, I owned my mistakes, and we implemented new programs in our organization to crack down on it."

7.4 Changing MPA Management

In Malaysia, 78 percent of stakeholders felt that management strategies that failed could not be changed. On the other hand, 77 percent of Indonesian said changes to management could happen. Most Malaysian stakeholders, even those in Marine Parks, felt that the large bureaucracy was the reason that change was slow or non-existent. Indonesian stakeholders felt that regular, face-to-face meetings as well as the localized management scale made change easy. They did note, however, that there were still problems related to graft and corruption at the local scale that could impede change but that localizing management meant that there were less people to have to bribe.

7.4.1 Malaysia: Bureaucratic obstacles to change

Marine Parks officers and other stakeholders cited the complex bureaucracy as a reason that adjusting management strategies is difficult. Stakeholders working in conservation or in the dive industry also cited a lack of capacity and will from the Marine Parks officers as another reason that management could not easily change. Take, for example, the scientific monitoring programs that occur in the park, largely through third-party agreements with local universities. In an interview with a faculty member leading coral species diversity, rugosity and fish count surveys with his students, I asked whether or not this data could be used to change management strategy in the parks. The faculty member replied that any data that they find is not used in order to assess current management, or even to recommend new directions in management, but is simply filed with Marine Parks. When I pressed for a reason why the science itself cannot inform management, the faculty member suggested that fear of losing jobs may play a major role, where if a policy needs to be changed, then somebody must shoulder the blame for the previous policy that, by nature of it needing to be changed, was not working. Other respondents working in NGOs and in conservation could list dozens of suggestions they had made to the central offices of the Marine Parks, only to be stalled for months and ultimately ignored.

If you find evidence for example that ships just outside of the park are dumping bilge water and causing fish die offs, you can file a report from your NGO to Marine Parks. They will tell you that since it is happening out of the park itself it needs to go to another agency, if something happens on land, it is up for debate whether it falls under the state government or the Marine Parks. That's all to say that change can happen, but it could take years of following up on reports that you send to Kuala Lumpur to various agencies. It's all very muddled, and ultimately it comes down to how much time you have to spend chasing people down, only to be told you spent the last year with your report sitting on the wrong person's desk.

The bureaucracy blocking changes to management was a remarkably widespread view among informants, from both within and outside of the Marine Parks. Despite these trends, some stakeholders were working for change. Stopgap interventions in the face of ecological crises were present thanks to the efforts of a few dedicated NGOs, primarily Reef Check Malaysia. During this research, I observed that when the need for immediate or long-term changes in management policy presented themselves, stakeholders across sectors would immediately call a prominent member of Reef Check Malaysia, Alvin Chelliah, in order to address the problem. During this time, there was a need for policy change in solid waste removal from Tioman Island, which had fallen into the gray area of ministerial confusion over whose remit it was to remove waste. One stakeholder illustrated the confusion, "Was it the state of Pahang? Marine Parks? Other environmental agencies? Was it the responsibility of the resorts, which are responsible for most of the waste?" These questions dogged local stakeholders while garbage continued to pile up. Ultimately, this same conservation leader from Reef Check played a pivotal role and organized locals in the village to come together and plan and implement the financing and removal of solid waste from the island.

Similar confusion over which government agency holds responsibility is the case of sewage removal. "Nearly every hotel on the island uses a septic tank. These haven't been emptied since these places were built. We need to address this problem before these things overflow and enter coastal waters," said a professor at the University of Teknologi Malaysia (UTM). Another respondent mentioned how it has been a major challenge for locals to figure out what government agency is responsible: "On the mainland, the question would be easy to answer, but since we are on an island, these are more complicated, and fall under several competing jurisdictions depending on what the issue is. In the case of sewage, nobody wants to deal with it." The need for change to waste management policy on Tioman is beginning to become a widely known issue, but many on the island expressed the perception that the same individual who played a major role in solid waste organizing will take the lead in the sewage issue. "Without him, nothing would get done," said a prominent man in the Tekek village.

Another example of the need for policy change discussed by stakeholders involved rule breaking by individual divers. Currently, the Marine Parks engage in information sharing to combat these problems. Throughout the park, from restaurants, to hotels, to the boats that ferry tourists to and from the island, are posters that spell out the rules for divers and swimmers in the park. First and foremost is that no removing of wildlife is allowed and secondly that no touching of the corals or organisms on the reef is allowed. Several stakeholders suggested the need for spot checks by Marine Parks officers, who should survey guests and evaluate whether or not local dive shops encourage their clients to follow the rules. "There is a need for more oversight on the part of Marine Parks. Currently they do nothing to make sure rules are enforced by local dive shops and the reefs are worse than they have been in 20 years because of it." Another stakeholder said the following, "Many of us have asked for years for greater enforcement on the reef themselves, but it is impossible to change." Frequently, respondents used words such as impossible, hopeless and slow when describing the ability to change tack in Malaysian MPAs.

My own observations from over 25 dives in Malaysian MPAs found hundreds of examples of rule breaking, often in plain sight of instructors or divemasters and sometimes with their added assistance. Occasionally, a few rule-breaking shops would feed fish for their clients' photos, or touch corals in order to bring an organism into better view for their clients. Several stakeholders suggested a real need for changes to enforcement, but also acknowledged quickly that such change is nearly impossible in the Malaysian Marine Park system. "Who would we even discuss the need for change with?" one seasoned veteran of the Perhentian Islands dive industry asked. "Nobody is even in their office to ask," he continued. "Besides, people in the shops can easily rattle off a list of names of those shops and divemasters who are constantly breaking rules. We see it every day. We call them, 'repeat offenders' and there's nothing we can do."

Many respondents emphasized that although stakeholders may widely recognize management problems, often Marine Parks officers do not recognize these same problems. "It is not clear to me whether Marine Parks officers have the training to recognize that the widespread touching, trampling, and handling of corals is even a problem, so why would they want to change it?" said one stakeholder working in conservation. Many more echoed these sentiments. "We aren't really sure if many of the Marine Parks employees can swim, let alone recognize naughty behavior on the reefs," said another stakeholder working in the dive industry. He added, "They aren't all completely ineffective, but a great deal of them are. You have great ones here and there, ones who care, but the majority of them cannot even swim". Similarly, another interviewee said the following, "Just because we see touching and trampling every day on the job does not mean that Marine Parks knows it is a problem. I have reported the extent and scale of the problem again and again," he said, referencing his years working in diving and conservation, "but when you report issues to Marine Parks, it just gets lost in the pile. I have invited those guys again and again to come out and see the problem, but nobody has ever accepted." One stakeholder summed up the problem by saying the following:

> Changing management strategies would imply that they were managing to begin with, which we have already established in this conversation that they are not. But let's say they were. You see people out there every day walking across the corals on the reefs right next to shore. You can literally see them from the main offices of Marine Parks and nobody does anything. That is why there is a perception on the islands that the officers do not do anything, because even the most bold violations of park rules are not stopped. Now, if they can't be bothered to walk down the beach and ask someone to get off the reef, why would they go diving and check to make sure people aren't trampling and touching under water. No way.

7.4.2 Indonesia: Change is possible—but not without graft

Indonesian MPA stakeholders had very different perceptions on the ability to change management. In Indonesian sites, 77 percent of stakeholders agreed that the reef management institution can change its programs when it becomes clear that they are not working. Stakeholders commonly mentioned two explanations for the ease in changing management strategy: the local scale of management responsibility allowing for rapid action as well as the frequency of meetings held within institutions, which provides the setting where problems can be addressed.

Respondents across Indonesian field sites nearly universally mentioned the local management scale as the main reason for the ability to change management policy, rules, or strategy. For example, stakeholders in Lovina spoke of time periods when certain areas of reef were so heavily damaged that all access, including line and pole fishing, which is allowed for locals on all reefs, had to be eliminated. According to a representative interview, "Once we saw this area begin to come back, and the coral to look like it did five years before, we eventually opened it back up, and let people fish the area again with line and pole." Another stakeholder added, "If we were to have to go to local government, or even worse, Jakarta and ask to reopen the site, we would have to wait years and pay a lot of bribes." Another similar management strategy discussed in the case of Pemuteran was ending the ability to fish with gill nets on the reefs in the early days of the reef management organization. Stakeholders remembered how it became known, thanks to several people in the area with backgrounds in conservation, that net fishing was causing significant damage to the reefs. Eventually, "we held several meetings and decided that only fishing with a line was allowed and that net fishing had to end on the reefs. We didn't need to wait years for Jakarta to okay this." The time-consuming nature of having to "go through Jakarta" was frequently compared to the ease at which management changes occur in the co-managed institutions.

Many stakeholders attributed the ease of policy change to close familial ties within the institutions. In the representative words of one stakeholder, "Every family in the village has a member in the [MPA organization]. This means that the people believe in what it is doing, and if it makes rules, they follow, even if they don't care about the reef." Others echoed similar sentiments, such as a boatman in Pemuteran who said, "The original reef closures were not popular among fishermen, especially older ones. But family members, neighbors, and friends who were positive about the closure persuaded those who were not." The ability of strong social ties to persuade people to agree to changes in management policy was mentioned repeatedly in interviews using words such as *our village, family, close, relationships*, and *working together*.

In addition to social ties, stakeholders also mentioned the frequent face-to-face meetings with MPA members as another reason that management changes could be enacted. Monthly meetings, such as those of the reef management institutions of the villages of Lovina, were the main forums where members could bring up and discuss issues to be addressed by voting. Stakeholders repeatedly mentioned this process as one of the best tools that the organization presented for changing bad management strategies or enacting new ones. They also praised the monthly meetings as a pathway to problem-solving at the institutional level or among individuals who had disagreements relevant to management.

Many respondents mentioned that local management, although effective, had its problems. Even local management was still subject to delays and inefficiencies caused by corruption. Corruption was a widely discussed topic among all Indonesian respondents. "Indonesian government is corrupt," said one local leader. "There is no ambiguity about this, it is widespread across Indonesia, and it is just a fact of day to day life here. But, when we can manage locally, we pay off people less and less." This sentiment was widely echoed, where interviewees mentioned the ability to make positive changes to management only existed because many layers of bribes and kickbacks did exist, and instead there were only a few. "You can buy anything in Indonesia. You can buy a PhD degree if you know the right guy," joked one respondent. "If you say 'I want to ban net fishing in my village' and there are well connected people who do net fishing, then you might be in trouble because there would be a lot of people to pay." The respondent continued to explain, "But because we manage locally here, well connectedness doesn't mean as much when some people may lose out due to certain rules."

Respondents discussed several ways that corruption could delay management changes. Bribes, local infighting among well-connected people at all scales of government and jealousy over attention, such as international environmental awards, were reasons that smart policies could be blocked. Even though local management cut through these obstacles significantly, there was no way to avoid them altogether. One local hotel owner in Pemuteran, whose name is widely known in global environmental circles for his exhaustive work in reef conservation, said, "In the early days especially I had to bribe every living soul to set up my resort. It doesn't matter that I have community programs, that we actively fund much of the conservation here, and so on, the bribes needed to be paid." Another long-time conservation leader who owned a dive center said the same, "If I pocketed every cent I made, I would have made twice what I have earned to this day. It all goes to bribes. You get an article written about you in the paper for conservation? The head man comes to you for a bribe."

There was a widely held perception among respondents that corruption varied with the quality of the village leadership at the time, which is different from the leadership of the MPAs. The head of the village could be very open to conservation, or he could be a jealous personality that demands payment when any national attention or spotlight is placed on the community. The latter type of village leader could also impose arbitrary permitting requirements on local conservation actors that served as a major disincentive to assisting the ecological work of the reef management institutions. This is a serious problem in Indonesia because the process where outside NGOs partner with local reef management organizations is where much of the scientific monitoring comes from on these sites. Every reef or dive site with access from the shore in this study had signage with a variety of international NGOs such as Conservation International or Reef Check Indonesia. These NGO partnerships with MPA managers were key for designing management strategies, from larger-scale visions and branding of areas for tourism purposes to the specific rules that dictate terms of access.

Rules could include, for example, who can and cannot access the reef sites for business purposes, such as in Tulamben where individual dive companies had negotiated tariffs with the local village MPA in order to access the reefs and had to abide by the Tulamben-specific rules. Other rules could protect Indonesian jobs from foreigners. For example, some sites would only allow divemasters from an Indonesian background. Interviews with NGO workers suggested that the support they give to village MPAs was not only critical but also a constant struggle because of the bribes and backroom deals that have to be struck. "Sometimes you need to pay certain elders in the village, even though I am from this village myself, and now work for an NGO that is surveying their reefs for them for free. It isn't fair, but it is Indonesia," one respondent remarked when preparing himself to conduct a dive survey on a local reef. "Local management is better because we don't deal with bribing everyone all the way up to Jakarta, but you need to pay people off locally if you want to help." Another NGO worker noted, "Local leaders are often older. You don't see so many young leaders taking bribes like this, it's an old way of life. They see that we are working for a big NGO and they want a payment, it doesn't matter that I am Indonesian."

Other stakeholders spoke of their role within reef management organizations and how they tried to start side programs only to be sidelined by bribery. One conservation leader described a project he set up involving amnesties for turtle eggs and hatchlings:

Back in the 1990s and 2000s we set up this turtle program to incentivize people not to harvest eggs or keep hatchlings. Everyone wanted their cut, the village head at the time reported me to his friend in local government who came demanding permits, which here is code for a bribe. I refused to pay for a while, and people kept coming back and coming back asking for permits and bribes. Finally it worked out, but it took a lot of payments in order to offer people a financial incentive not to cook sea turtle eggs or the hatchlings.

Stakeholders from expatriate backgrounds who worked in the dive sector echoed the complaints on graft, saying that even though their businesses were foreign investment, and they volunteered extensively for conservation, they were sometimes the most heavily penalized. Indonesian stakeholders in village sites generally agreed with the perceptions of expatriate dive shop owners, viewing them as frequent funders of conservation initiatives, volunteers or sources of assistance to the main reef management institution. For example, several dive shops would frequently lend their boats for patrols for the MPA or for larger publicity events. Expatriate dive shop owners expressed significant frustration over the amount of graft. In the words of one representative respondent: "I have been in this community for 25 years. I only hire locals, and pay them a fair wage. I am constantly hassled by the village leadership or friends of villagers in local government to pay bribes and 'permits' which are bribes."

The discussion often focused on the variability of the intensity of graft that depended on whether or not a given village head was sympathetic to reef conservation or not. "If the head of the village is pro-conservation, I get minimally harassed for bribes to keep my eco-friendly shop open. If he is anti-conservation, well times are hard." Many had been in these villages for decades, watching them change from little-known spots well off the beaten path to globally known destinations for diving enthusiasts. "I was here before there were ATMs," said one owner responsible for many reef conservation programs in his community. "I was here before they formed their reef management organization, and we used to use guns to scare poachers off the reefs." He continued, "I won't be here for much longer. The constant bribery, the corruption is getting to me. Reef management is good, it could be great if corruption did not impede good work for grudges." The grudges this respondent referred to included, for example, times when people who did not like him were elected into the village leadership and he had to pay more fees to operate his business than during times when people who did like him held power.

Grudges mentioned by this particular stakeholder were another common topic that came up in interviews when respondents discussed the problems with graft. "Indonesian village culture has these grudges, you see them it in small places all over the world, everyone knows everyone and everyone talks. Sometimes though, it stops well meaning people working to save the reef." Stakeholders specifically mentioned stories of times when businesses were denied permits to engage in conservation work because of inter-village rivalries and grudges. Other stakeholders mentioned times when they had won awards for their work in reef conservation from globally recognized organizations such as the United Nations Development Programme (UNDP) only to be denied the use of a boat by a hotel owner next door because he was not invited into the award photo opportunity. "Even though we do not have to deal with the bureaucrats in Jakarta," one stakeholder remarked, "sometimes the village itself can be just as much of a snare if you want to get something done."

A large amount of stakeholders worried that conservation itself was taking on a competitive attribute in the villages, where people saw that money and international accolades come from conservation work. Thus, some used connections in the village leadership or in the local government to block the conservation work of others, lest the others gain international awards, funding or press for the work. Although stakeholders did mention social ties repeatedly as a reason that management changes were possible, the story was not always a positive one, whereby the same social ties could also impede management change.

7.5 Innovation

There were significant differences between Indonesian and Malaysian MPA stakeholders involving innovation. I asked stakeholders whether the MPA allows businesses or other stakeholders to engage in innovative activities. Innovative activities could include dealing with ecological crises, contributing to conservation or engaging in win-win activities that promote conservation and business. In Indonesia, nearly all stakeholders interviewed said that this was a possibility (95 percent) and the field sites had many vibrant examples to show. Malaysian MPAs were nearly universally perceived to not support innovative activities for conservation.

7.5.1 Indonesia: Widespread innovation

In all of the Indonesian MPAs, innovative programs geared towards enhancing both conservation aims and tourism were present. These were often the main attractions in the villages. A representative case of innovative interventions for conservation is the BioRock formations in Pemuteran, a partnership between local tourism elites (TEs), an international NGO and academics to install charged wires underwater that they claim enhances the growth rate of coral reefs. These sites attracted people from all over Bali to make the long journey, four to six hours from the main tourist cities. I observed BioRock formations a half-dozen times through diving and snorkeling and was struck by their beauty. The team that installed them engaged the local community in their design, installing Biorock frames shaped like a giant sea turtle for example. Tourists flocked to these sites and MPA managers charged every visitor the equivalent of $2 to visit any Biorock site. This fee paid for the electricity-generating solar panels installed on the formations and paid locals for maintenance work. Informal interviews with 10 tourists on the beaches of Pemuteran led me to conclude that these features were a major draw for tourists. Tourists also liked the educational focus of Pemuteran, where nearly every resort, office or restaurant on the Bay had information about the BioRock formations, coral reef ecology or village-based coral reef management. One stakeholder summed up the importance of projects such as BioRock to the community:

> Education, conservation, and tourism is the three part business plan for our community. You cannot have one without the other two. Projects like BioRock bring attention and visitors. It shows the world that our small village is doing big things for conservation. We know it is good for the environment, but we also do it because we see that it has a big effect on our incomes.

Another stakeholder reflected on the catalytic potential of projects such as Biorock, noting that before Biorock, people did not think that conservation projects could bring in tourists. Once people began to see that they could, more and more conservation projects began to take root in the village.

> Before BioRock, local people did not care as much as they do now. When people started to see that you can build something like BioRock on the reefs and people would pay us to come and look at it, people became interested. Smaller projects started to grow from the BioRock Project, like the name plate project where you can use the technology to grow corals on a metal wire that spells your name. This means people pay us money to grow more coral, and we love to do it.

Another respondent spoke on the scientific realities of such programs, noting, "I know there is a lot of talk in the conservation community about the inability of these projects to scale up, but we have whole villages investing in conservation, instead of, say, fishing. That is huge." He concluded by saying, "If you had asked me 25 years ago if this was possible, I would have said no. These reefs are history. These projects made people see that conservation and incomes are one and the same."

Respondents were asked whether MPAs supported innovation. Respondents generally felt that MPA managers and innovative conservation projects came from the same set of people. Stakeholders in Pemuteran spoke of MPA managers and BioRock Projects as one and the same. "A lot of the guys who work as Pecelan Laut also work on BioRock stuff. This is a small place. People who monitor and protect the corals are also the ones running these programs, so yes there is a lot of support." Other respondents suggested that without the support of MPA managers, these projects could not happen. In the words of another respondent, "it's the Pecelan Laut guys that are out there everyday, and the BioRock guys are typically Pecelan Laut. They do all the diving, maintenance, and work on the projects. Sometimes for free."

A second example of an innovative conservation project supported by the MPA is Amed's underwater temple structures and its artificial reef-building program (Figure 7.2). Village leaders and MPA managers as well as several TEs worked together to install a series of Hindu religious statues in the Jemeluk Bay area. These "underwater gods" as they were called were a major tourist attraction for both snorkelers and divers. In a half-dozen informal interviews with tourists on the beach, people generally came to Amed for the diving not knowing about these statues. But they found the statues to be one of the best dive sites in the village, because they blended the beauty of the reefs with Balinese Hindu culture, two features that bring many tourists to the islands. The main creator of the underwater temple statues also owns a dive resort

Figure 7.2 Underwater sculptures. Top left: Underwater sculpture of Hanuman from the Hindu epic Ramayana. Top right: Artificial reef in Amed installed by a hotel and restaurant owner. Bottom left: Launching ceremony for artificial reef structure on Independence Day. Bottom right: Artificial reef installed by a resort in 2014.

in the village that is well known for its persistent efforts in reef conservation. He said that investing in projects, even though it may cost him money, pays for itself very quickly. "People come to see us, people stay in my hotel. More fish are in the bay, and word travels that we have a lot of fish to see in the beautiful setting of the temple." Respondents noted that in years where the seasonal storms break a lot of coral on the reefs, artificial underwater attractions continue to bring visitors to Amed. "One year we had really bad storms and a lot of coral was broken. The reef was ugly, but I still brought clients to see the underwater gods, and many liked those and wanted to see them a second time."

When asked whether the MPA managers support such projects, responses echoed those of Pemuteran. Most of the people who manage the MPA are those who fund, implement and maintain these types of projects. On Indonesian Independence Day in the summer of 2015, I witnessed the sinking of a new artificial reef structure. This structure was a type of underwater sculpture that celebrated local religious customs. In the words of a representative respondent, "religion is important to us, and projects like these link our religion and our livelihoods." This particular stakeholder had worked in the dive industry for over a decade. He noted that without local TEs and MPA leaders, these projects could not occur. "Local leaders in the village pay for these projects, and then they also conserve the reef. We are very lucky that people here think that protecting the coral and business are the same thing."

Pemuteran, like Amed, also has its own underwater temple complex installed by a conservation leader, TE and award-winning conservationist. Many in Pemuteran felt that this project was a great "back up plan" in case reefs became diseased or large storms impacted tourism. In the words of one stakeholder, "When a big storm comes, you can see its impact on the reef. Word gets around, and we can have a slow year. If people come and see the underwater temple though, that's different." Another stakeholder echoed this perception, "Maybe we have a bad year environmentally, where a lot of the coral gets broken in a storm. Because of local leaders' work on the reef, people are still coming and our hotels and restaurants are full." A third respondent added, "Pemuteran is lucky to have people who invest in things like Temple Gardens, it brings in tourists and it shows local people who may not know that people come here to see healthy reefs and to watch us look after our reefs. People like it."

7.5.2 Malaysia: Fear of fragmented responses

Malaysian stakeholders, when asked about innovative projects that link conservation, generally felt they were impossible. "I have been trying for years to put an artificial reef out there in front of our shop. There is no way it will ever be approved by Marine Parks." Similarly, other stakeholders remarked that individual efforts towards conservation could benefit from them when there was a bleaching outbreak, disease outbreak or period with a lot of storm damage. "Artificial reefs are sort of trendy now," said one dive shop owner. "They won't save reefs, but they bring fish around and we need fish. We couldn't put these in even if we wanted to."

Malaysian Marine Parks officers were skeptical about the potential for individual innovations to make gains in conservation for two reasons. First, they felt that if everyone was building their own projects, this might actually damage the reef. Second, they thought that supporting individual interventions in the system by stakeholders and businesses was not part of their job description and could result in their termination from Marine Parks.

Many Malaysian Marine Parks stakeholders felt that supporting TEs, NGOs or other stakeholders in the village intervening for conservation was a bad idea that would lead to a "free for all" where projects may become abandoned and possibly damage the reef. "If we allowed locals to come in and build structures under water, storms could damage them and drag them across real living reefs." Other Marine Parks officers held this same view: "If resorts could build their own reefs, and monetize access to the reefs, visitors may get angry that they had already paid an entrance fee to visit Malaysian Marine Parks and are now being charged again." Others working in Marine Parks felt that they did not have enough personnel to make sure that these projects were compatible with the aims of the MPA. The MPA was meant to protect the reefs and not as a place where statues or artificial reefs need to be built. Some Marine Parks officers pointed out that they had placed artificial reef structures near the Jetty in Tioman to respond to the demand for these types of projects. The result was an increase in fish around the jetty area, which draws tourists. Interviews with several tourists suggested that these artificial reefs were an attraction for them because they could not swim, but given the shallow depth and the clear water, they could see fish from the jetty.

Marine Parks officers generally thought that supporting community interventions on the reef ecosystem was not a part of their job. Many felt that it was their job to build these types of things and that was why they collected an entrance fee to the park. "We paid for these [artificial reef] structures and put them into place. The community or dive shops did not need to be involved, that is why we collect money, to do projects like these," said one seasoned Marine Parks officer. He added, "We want them to take people diving, not to have to install these reef structures." Others voiced concerns that if they did allow individual dive shops or NGOs to have a creative license over conservation programs on the reef, they could lose their jobs. "If I let a hotel put statues near the reef, and these become loose and break corals, then I am accountable for this damage," said a Marine Parks officer. "It is hard enough and expensive enough to monitor the reef, but monitoring all of the community and all of their projects would be impossible for us."

7.6 Summary and Conclusions

This section found adaptive capacity to be higher in co-managed communities. Co-managed Indonesian case sites mentioned man-made environmental problems, as well as strategies and programs employed to fight these problems. In contrast, Malaysian MPA stakeholders mentioned similar problems, but were at a loss for ways and means to end them. Indonesian case sites had large-scale ecological crises such as crown-of-thorns outbreaks. While locals knew they could not stop these outbreaks, they did form programs to eradicate some of the organisms. Such programs are difficult to scale in order to completely eliminate the starfish, but this program and similar programs amplified the pride that the community took in managing its reef. Malaysian MPA stakeholders mentioned similar ecological processes, such as coral damage during the monsoon season, but expressed an inability to close off damaged areas to divers and snorkelers in order to allow recovery. This inability was a combination of Marine Parks officers fearing job loss when they make decisions that may be unpopular and local-level discrepancies between different stakeholder groups in terms of their knowledge on marine ecosystems. In other words, poorer, less-educated stakeholders from local villages would be against this while wealthier stakeholders from the mainland would support it.

Learning is the key to adaptive capacity in case sites, and my data showed that learning was a function of the flat management structure in the MPA, local pride in coral reef conservation efforts and clarity of responsibilities. The barriers to adaptive capacity in these field sites included bureaucratic red tape, fear of losing one's job, corruption and small-town interpersonal rivalries or jealousies.

Finally, innovation was alive and well in Indonesian co-managed MPAs because local business owners had the freedom to invest in such projects. These projects, such as underwater temples combined conservation (the need for fish habitat after coral cover loss from storms) and tourism (these sites are very popular with dive tourists). Because of the horizontal scale of management, such projects were well funded, common and highly admired by visitors and locals. In Malaysian MPAs, many business owners expressed an interest in investing in similar projects but feared the arbitrary permitting process in the MPA. A critical part of adaptive capacity involves expanding the possibilities that stakeholders have to respond to crises. While there was evidence for creative interventions such as artificial reef structures in Indonesia, Marine Parks Malaysia felt that allowing these would result in fragmented projects. Indonesia must combat corruption at the local level in order to realize its full potential in adaptive capacity. Malaysia can begin to improve adaptive capacity of its MPAs through genuine bureaucratic reform that streamlines permit applications from NGOs, for instance, and makes reporting crises easier for stakeholders.

Chapter 8

POLICY RECOMMENDATIONS FOR MARINE PROTECTED AREA MANAGEMENT IN DEVELOPING COUNTRIES

8.1 Overview

In this concluding chapter, I summarize the lessons learned about integrated management, legitimacy and adaptive capacity from the marine protected areas (MPAs) described in this book. Then, I offer policy recommendations for Indonesian and Malaysian decision makers with responsibility for managing coastal biodiversity conservation. I conclude with recommendations for tropical coastal biodiversity conservation in general, arguing strongly on behalf of co-managed institutions for natural resource management and highlighting what I argue are the strongest explanatory factors for the success of Balinese MPAs: stakeholder engagement that spans economic sectors, gender, religious beliefs, socioeconomic class and education levels.

8.2 Insights on Integrated Management of MPAs

Integrated management of MPAs occurs when economics, conservation and social systems are given consideration before any management decisions are made. MPA managers are more likely to see success when they consult stakeholders about decisions that might impact reef tourism businesses, such as closing parts of a reef impacted by a storm or bleaching damage, or levying fees or regulations on those who depend on reef access for a living. I found that integrated management (when local business and MPA managers work together) is present in co-managed Indonesian MPAs and absent in Malaysian Marine Parks.

In Indonesia, this is advantageous to stakeholders such as the dive industry or boatmen who bring tourists to snorkel because their voices are heard before restrictive rules are enacted, often eliminating the need for them to break the rules. Integrated management acts as a "steam release" where potentially harmful activities are not eliminated outright, but instead limited or restricted. For instance, isolated instances of line and pole fishing where fisherman take only what they can carry were incorporated into the rules for several Balinese MPAs reviewed in this study. In Malaysia by contrast, the lack of integrated management, and the lack of stakeholder engagement overall, resulted in the loss of employment in fishing for nearly every resident in every village within MPAs in the 1980s. This created an atmosphere in which rule breaking was tolerated, ignored or even seen as a survival tactic among village residents and some Marine Parks officers. This legacy persists today among residents who generally do not feel connected to the reefs for their livelihoods through tourism.

8.2.1 Linking economics and conservation

In my view, Malaysia needs to strengthen the links between businesses that benefit from reefs and officials who enforce the MPAs. As I demonstrated in the previous chapters, dive industry stakeholders believe that Marine Parks officers do very little in enforcement and block meaningful conservation projects such as artificial reefs or scientific research. Likewise, many dive industry stakeholders resent being asked to perform maintenance on park facilities, such as mooring points or removing debris from the reef, when they see so little activity from the Marine Parks officers themselves. There is a need to ensure dialogue between Marine Parks officials at all levels and the dive industry sector. This is critical because the dive industry is a major source of economic activity in the MPA. In addition, many who work in this sector have marine science training or other capabilities that would be helpful to Marine Parks. Were this relationship to be repaired, there would be increased possibilities for collaboration on cleanups and engagement programs for both local residents and tourists. A simple example could be a monthly meeting at the Marine Park office where officers and dive industry workers raise problems, pose solutions and decide on the best ways of solving problems. There is a widely held perception among dive industry stakeholders that Marine Parks officers serve no purpose. These meetings and conversations could repair these relationships, but they would need to be long-term, enduring and genuine places where discussions on the sustainability of the MPA can occur.

A second way to improve the relationship between dive industry stakeholders and MPA officers would be to end the arbitrary, confusing, expensive and lengthy process of applying for permits to do research or build conservation projects in the MPA. Malaysian Marine Parks should not deny qualified conservationists permits to do research or train divers one day and then ask for their help in removing garbage from the reef on another day. The permitting system needs to be transparent, and conservation projects such as artificial reefs should be viewed as an opportunity to increase tourist and local interest and education in coral reef science. As was seen in the Balinese example, when tourists and residents feel like they are "helping" or taking part in conservation (whether or not that is the reality in artificial reef building is a subject of much debate), their interest in conservation grows. As it stands, stakeholders who could otherwise serve as valuable allies to Marine Parks—experienced scientists, PhD students, locally based NGOs—see applications sent into a Byzantine bureaucracy and then turned down with little to no explanation. I suggest that higher-level Marine Parks officials should allow local officers to approve proposals. Local-level officers should view such proposals as (1) a way to engage business stakeholders, (2) a way to engage academics and (3) a way to build the capacity of Marine Parks in areas where they are weak.

8.2.2 Trust of business

There is also a need for Malaysian Marine Parks to win the trust of business owners who feel that Malaysian Marine Parks officers do not do their jobs and believe that the officers play a part in denying permits to qualified conservation leaders. In order for Malaysian Marine Parks to earn the trust of local businesses, they must find a way to end the most flagrant and serious abuses of MPA rules, namely, off-season illegal fishing. Dive industry stakeholders cited this activity repeatedly as the key piece of evidence proving that Malaysian Marine Parks are ineffective. There is no reason that a predictable and widespread activity that breaks the most important MPA rules should occur simultaneously across peninsular Malaysian MPAs every year. What is needed is an aggressive publicity campaign that will bring this annual problem to the public's attention and a push from the federal government to stop illegal fishing.

Indonesia's flashy new campaign, which involves seizing and blowing up illegal fishing vessels, is very popular among its citizens as a sign of national strength by the Joko Widodo administration. Malaysia's government could conduct a similar campaign, making examples of a few frequent violators. This would help to win over public opinion. Graft must also be tackled, since many respondents said that illegal fishing occurs because owners of the vessels involved are well connected and wealthy. If Marine Parks were to implement a year-long campaign aimed at tamping down on this predictable, visible and widely loathed activity, it could win back some of the goodwill lost among numerous stakeholders.

In addition to ending gross violations of illegal fishing in the offseason, dive industry stakeholders in both Indonesia and Malaysia must do a better job of ensuring that the dive tourism economy is sustainable. Two prime culprits need to be addressed. The first is the mixed incentive for shop owners who are pressured to issue a certain quota of dive certifications (by certifying agencies such as PADI or SSI).[1] This results in certifications being issued to people who cannot dive without causing serious damage to reefs because they cannot swim and are using their buoyancy vests as a means of staying afloat. The consequence in the long run is that damage accrues to the same reefs that attracted people seeking to earn dive licenses in the first place. Dive certification organizations such as PADI and SSI, as well as their licensed dive shops throughout Indonesia and Malaysia, need to work harder to prevent licenses from being issued to people who cannot swim and cause damage to the reefs. Dive operators and certification organizations tend to operate in unaccountable ways in developing countries in Southeast Asia. This requires urgent reform.

Some nongovernmental organizations (NGOs) exist specifically to lessen the impacts of non-accountable dive companies and to ensure that employees use best environmental protection practices when training future divers. Prominent examples include the organization Green Fins or PADI's Project Aware, both of which were in operation at several of the field sites. It is important to note, however, that even in shops working closely with these organizations, dozens of instances of behaviors by instructors and clients alike that damage reefs (such as kicking coral) were observed.

In a majority of dive shops visited for writing this book, lower-level employees (such as divemasters) allowed clients to touch corals, hit corals while taking photos or feed fish. Higher-level employees, such as shop owners and dive instructors, spoke at great length about how these behaviors were not allowed in their shop. A striking example occurred in one of the most famous dive centers in Indonesia, owned by a true leader in conservation, where I conducted a multi-hour interview about the many initiatives undertaken by this shop to spread the message of conservation. Later that day, while on a dive with one of their divemasters, I watched as he repeatedly allowed a couple (from a Western country) to touch plating corals while taking photographs. This divemaster spoke almost no English. When I asked him in Indonesian why he allowed this to happen, he replied that he did not want to make trouble for the clients because they might give negative feedback to his supervisor. Thus, the impact of power differentials in dive centers must be addressed: lower-level employees such as divemasters have an incentive to ignore diver misbehavior

---

1 PADI stands for Professional Association of Diving Instructors and SSI for Scuba Schools International. They are the largest dive certificate-issuing organizations in Southeast Asia.

lest it cost them their jobs (which are already low-paying and insecure relative to similar positions in the West). This must be addressed within individual dive centers. Owners and senior employees need to engage lower-level employees and encourage them to be more strict with clients who interfere with corals and other wildlife.

This is also an area where NGOs such as Green Fins and Project Aware can help. Many dive shop owners perceive these organizations as "green-washed black holes for owners to throw money into," in the words of a Malaysian shop owner. There was a widely held sentiment that training or information provided by these programs had limited use for employees, or else they included information that was already known to employees. Some stakeholders suggested that money spent on programs such as these could be better spent helping lower-level employees improve their English communication skills so they can correct the behavior of tourists who damage corals. A dive guide in Malaysian Marine Parks, who did not stop her divers from touching corals, explained in Malaysian that her English is not "good enough to explain what they should not do."

8.2.3 Too many cooks in the kitchen: The future of coral-focused NGOs

In the previous chapters, I mentioned that at both Indonesian and Malaysian field sites, there was a worrying lack of science to inform management. That is not to say that there are no scientists studying the coral reefs. Quite the contrary. There were many scientists from a range of local and international universities studying reefs at each field site. I met research teams at every field site through sheer coincidence. What is a problem, however, is that MPA decision makers and managers perceive a lack of scientific data they can use to inform their decisions. Or in some cases they are confused about what to do with the data they have.

For example, in a community where a major Australian university was conducting bleaching work through citizen science training, many dive guides were unclear about how they could stop bleaching by reporting it online. Many stakeholders also felt discouraged, citing large-scale environmental processes such as climate change as a reason they failed to see importance of collecting data about their reef systems. "We could have all the data in the world," said one MPA manager in Lovina, "and yet we still cannot stop the bleaching. So, why do we have to keep this information? For scientists?" Other frequent criticisms of NGOs included the fact that there were too many projects and programs, many of which were short-lived, to keep track of. "The best ones are run by local people [like Iwan]," said a dive industry stakeholder in Amed referring to the dedicated work of a Conservation International leader, Iwan Dewatarma. This same respondent noted that programs come and go. Often the local community is left with fatigue from too many short-term conservation programs. Stakeholders repeatedly suggested that scientific studies need to engage in greater partnership with locals and long-term residents instead of foreign, often Western, NGOs and universities.

Given the frustration felt by stakeholders, I recommend that local governments enhance their capacity to act as a central repository for the data and information collected trough scientific projects in their villages. One central holding place managed by the regency or provincial level of government for all the data on Balinese coral reefs, for example, could be a starting point for a larger effort aimed at long-term contingency planning for large-scale events such as climate change. Local MPA managers can and should be trained on how to access, interpret and use this data for decision making on their local reefs. This links to another policy recommendation for Indonesian cases that I discuss further below. Although community involvement and local level stakeholder engagement was extremely strong in Indonesia, the top-down component of co-management, wherein government assists and supports local communities, was largely absent. There must be a role for local government and federal government, serving as a single point for all available data on Indonesian coral reefs or as a financier for increasing local leaders' data literacy and application skills. Engaging in long-term planning for future climate impacts can be another important role.

8.3 The Legitimacy of MPAs

8.3.1 Enhancing Malaysian legitimacy using lessons from Indonesia

Malaysian stakeholders, primarily dive industry stakeholders and local residents, see their MPAs as illegitimate. Presumably, they want to repair this. A reform in high demand was that MPA employees need to be at work and accountability needs to be a ministry-level reform priority. Stakeholders frequently discussed ill-fated attempts by the Department of Marine Parks to improve the perceived legitimacy of the MPAs such as through the installation of biometric scanners for employees. Employees were expected to use these scanners to check in to work, and show those in ministry headquarters that they were indeed working. But this reform does not do enough to get to the core of the problem. This issue in Malaysian Marine Parks is not that employees are not in their offices; it is that they are not out on patrols making sure illegal activities do not happen on the reefs. One source of reform may be employing people who want to be out on the water, namely those who want to work in marine science instead of people who pass a general civil service exam and want to work in any government office. Creating graduate entry-level hiring programs with Malaysian universities known for their marine science departments (where top graduates are offered positions in Marine Parks) might be one way to attract dedicated employees who know the science behind marine conservation, and do not want to shirk responsibilities at work.

In Indonesian MPAs, by contrast, the defining characteristics of MPA managers when compared to Malaysian MPA counterparts, was enthusiasm and pride in their job. This was absent among many Malaysian Marine Parks respondents. There was a noticeable difference in the pride respondents had in the MPA-related job tasks in Indonesia compared to Malaysia. There was also a remarkable difference in the way that local village residents spoke of Indonesian MPA managers compared to Malaysian ones. In Indonesia, young locals wanted to grow up to be involved with the MPA, and many said that only the best and brightest in the village go on to lead the MPA or the fishermen's cooperative that manages the MPA. Residents in Malaysian villages had no enthusiasm for Marine Parks officers, describing them as outsiders or as work-shy people eager for a government paycheck. There must be an effort in Malaysian Marine Parks to show the community that its officers are dedicated to their work; a first step may be reforming the hiring process to bring in people dedicated to marine conservation.

8.3.2 The need for top-down action in Indonesia

Local and national government can and should develop a more systematic way that NGOs and communities can work together to perform scientific monitoring. This could be as simple as creating a single data portal for all ecological and socioeconomic data gathered by Indonesian and international academics and NGOs. These data could be used across governments for contingency planning. For example, what will happen to the hundreds of millions of Indonesians who depend on coastal ecosystems when sea levels begin to rise this century? What will happen to these people if reefs undergo bleaching so severe that recovery is impossible? There is already a regional body with national-level offices where such an effort could be concentrated, namely the Coral Triangle Initiative, CTI. For now, much work remains to be done to involve communities in ecological monitoring and support contingency planning. I recommend greater effort on the part of regional level, provincial level, and national level government and agencies in Indonesia and the rest of Southeast Asia to support communities in reef management.

8.3.3 Posters are not enough

In top-down MPAs such as Malaysian MPAs, information sharing is at the core of legitimacy. Information sharing is happening in Malaysia, and I would emphasize that in every hotel, restaurant and building on the islands in Malaysian MPAs there are Marine Parks posters instructing visitors on how to behave on the reefs. However, some stakeholders do not perceive this to be enough. In fact, these critics deny such posters even exist, or alternatively fail to notice them. There is a real need to enforce rules regarding the banned behaviors listed on the posters. Many stakeholders feel that hanging a poster is one thing, but actually monitoring and enforcing the rules is drastically more important and noticeably absent in Malaysian Marine Parks. I have already suggested underwater patrols in the parks. These do not need to be conducted every day, or even every week. They can be ad-hoc in the beginning, and occur whenever the MPA offices and personnel are available. In March 2016 in the East Malaysian state of Sabah in the Semporna MPA of Sidapan, the main authority for MPAs, Sabah Parks, grew concerned after a series of scandals involving dive operators "went viral" on social media. These involved touching and harassing sea turtles and reef fish. They responded by implementing an "underwater marshal" program (Rakyat Post 2016). Perhaps the MPAs of Peninsular Malaysia could rely on Reef Check Malaysia to help them. Reef Check Malaysia was a key partner in implementing this program in Semporna.

8.4 Insight on Adaptive Capacity of MPAs

The ability to intervene in an ecosystem for conservation purposes, monitor the effects of such interventions and make the necessary changes requires capable institutions. Adaptive capacity was present in Indonesian MPAs but not in Malaysian MPAs. Marine Parks officers view management primarily in terms of installing swimming platforms and new jetties. There is a real need, though, for biological monitoring of coral bleaching and diseases along with the effectiveness of land-based policies aimed to preventing runoff and sedimentation. Even in the Indonesian MPAs in this study where adaptive capacity is high, the emergence of large-scale global climate events will make capacity the operative word in adaptive management. Local communities, no matter how well equipped they are to deal with ecological crises, can only do so much to counter global ecological crises.

8.4.1 Revising management

There is an urgent need for the Malaysian government to redefine the duties of the department of Marine Parks. There needs to be a top-down push to turn Malaysian Marine Parks into a scientific adaptive management authority with its own in-house team of biologists, ecologists and social scientists familiar with tropical marine systems. Marine Parks needs employees who can assess ecological data, make decisions about the need to close parks and undertake necessary restoration activities. Marine Parks needs to hire graduates from excellent Malaysian universities such as the University of Malaya and its renowned marine science program to work in the MPAs. Malaysian Marine Parks, on the other hand, cannot continue its loose partnerships with universities and NGOs. The outcomes are scattershot. Many stakeholders resent what they see as arbitrary permitting decisions with very little transparency. If the auspices of Malaysian Marine Parks is shifted to a more science-based adaptive management–oriented agency, then bringing in NGOs and academics as more formal partners might be helpful. This would mean that Malaysian Marine Parks would fund more research and make its findings public and easy to locate via a central data repository.

The most pressing situations that require reform in Malaysian Marine Parks involve even lower-hanging fruit. There should be an immediate effort to prevent tourists from walking on coral reefs in shallow areas in front of Marine Parks. There should also be an immediate end to highly visible poaching at night and poaching in the off-season. Septic systems and illegal runoff from hotels require urgent attention. Some of these activities occur so frequently and predictably that there is rampant stakeholder frustration that further impedes effective management. Much of this frustration occurs because many people know these illegal activities will happen, when they will happen and who is likely to be involved, yet nothing is done.

There is nothing short of a nation-wide challenge: a prominent and well-connected ruling elite is actively allowing behaviors that are unlawful. Respondents repeatedly cited illegal fishing in the MPAs by a company connected to the local sultan. Respondents would lower their voice to a whisper and ask me not to quote them when they relayed this information. There is a growing discontent with such corruption among the youth in Malaysia as evidenced by the *Bersih* Movement meaning "clean" in Malaysian. This social movement, which rose to prominence in summer of 2015, involved just under a hundred NGOs and thousands of protestors demanding an end to corruption in Malaysian politics (Rakyat Post 2015). Although their focus was on Prime Minister Najib Razak, and their core demands focused on ending corruption in politics, it applies equally to environmental politics in Malaysia.

8.4.2 Innovation cannot stave off global crises

Indonesian sites in this study offered multiple examples of situations in which businesses worked hand-in-hand with conservation leaders to raise the reputation of their villages as ecotourism destinations. Villagers, local hotels and dive shop owners in Amed built artificial reef structures shaped like marine animals and deities. A hotel and dive shop in Pemuteran installed an entire underwater "garden" of Balinese gods to demonstrate to visitors how passionate they are about preserving underwater beauty. There is something to be said about the positive impression that man-made underwater structures have on the hearts and minds of visitors. Showing people that humans can intervene in underwater spaces and create displays of beauty that parallel the drama of natural coral reef systems both excites tourists and locals and inspires an interest in what everyone can do to become more engaged in conservation. Balinese villages such as Amed and Pemuteran should be praised for their hard work in linking reef stakeholders, using those links to draw visitors and, in turn, boost the local economy.

Although innovation is a strong feature of Indonesian MPAs, larger-scale phenomena such as rising sea-surface temperatures and increased coral bleaching may end tourism in a way that artificial reef structures and underwater sculpture gardens cannot offset. Villages have come a long way from dynamiting and using cyanide to fish in the 1980s. They should be praised for making ecologically minded decisions, but looming larger-scale crises require immediate, urgent global action. Successful adaptation to a changing climate will require substantial emissions reductions as well as financing from Western countries. When rising seas inevitably ruin what villages have built, they will need help funding a new round of economic transitions.

8.5 Summary of Key Policy Recommendations for Indonesia

I offer three policy recommendations based on my research in Indonesia. First, in terms of governance, co-management is succeeding from the bottom up. But, it needs more support from the top down, specifically with regard to technical and scientific capacity-building as well as long-term planning for the impacts of climate change. Second, co-managed MPAs do an excellent job of engaging the community, but they need to focus more on formal, educational-based outreach to young people in order to create the next generation of homegrown marine science and policy experts. I recommend that multilateral agencies and NGOs target youth specifically through efforts such as scholarships in marine science or environmental social science. This generation has inherited a better life from their fishermen parents thanks to ecotourism. But the next generation can add to the rise of ecotourism as a new generation of scientific leaders in coral reef conservation. Finally and most importantly, corruption continues to plague Indonesian villages. It damages public will to make change and makes it harder to implement conservation programs.

8.5.1 Co-management needs greater support in government

Co-managed coral reefs require power-sharing between government and local resource user stakeholders. In Indonesia, MPAs are closer to what would be called "community-based" than they are to being "co-managed," meaning communities are largely on their own and they must fund conservation projects on their own. Government, both local and national, has played a role in attracting a handful of large, in-depth scientific studies, but there is a greater need to coordinate smaller NGO and university research programs. There should be one place where stakeholders can access data from all these programs online for free. There should be government support for providing scientific and technical reports to communities in *Bahasa Bali* (a remarkable number of locals do not speak or read *Bahasa Indonesia*) and there must be more training in communities on how leaders can view data and use it to make decisions about their reefs. There is a high number of international scientists and NGOs involved in scientific research on the reefs, but there is a worrying lack of local people leading and participating in these studies. Perhaps targeted investments in coral reef science scholarships at Bali's premier university, Udayana, could help create the next generation of local scientific leaders.

Most pressing is a pronounced need for more official institutional support from national and local governments for contingency planning regarding the risks posed by climate change. Increasingly in Indonesian MPAs, stakeholder concerns are shifting away from destructive fishing and focus more on bleaching events, coral disease and sea-level rise. Even robust local stakeholder engagement efforts do not stand a chance against rising seas. Government needs to support coastal communities and assist in their efforts to ensure the resilience of coral reefs.

8.5.2 The role of local youth

MPA leaders in Indonesian communities are doing an excellent job of educating and inspiring the next generation of conservation leaders. There is widespread pride in the men who constitute the MPA, with many younger men explaining their desire to join marine conservation efforts. Roles for women in the reef tourism sector are less visible although present in some villages, such as Amed, where each dive site and its fisherman's cooperative also ensures women's rights to serve as porters for dive equipment, thus providing them with a stable income. MPAs in Bali grew out of traditionally all-male fishermen cooperatives, which continue to be dominated by men. Going forward, engaging more young women in these villages could bring a new generation of workers into the conservation community. Since the youth across Balinese sites were so passionate about conserving marine resources, I suggest that NGOs or researchers look to invest in these young supporters, either though increased scientific training, scholarships or employment on conservation projects.

8.5.3 Ending graft

Corruption is one of the most significant practical barriers to stakeholders contributing meaningfully to conservation. Every time a respondent related a story to me of a project they tried to implement, there was always a follow-up detail about how many people they had to pay off to make it happen, and how often such payments had to occur before a project was finally allowed to proceed. Local as well as village-level governments in Bali must undergo serious reforms relating to graft in order to ensure that innovations originating with business owners working in ecotourism can continue. Many were tired of corruption and payoffs, and spoke longingly of finding another place in Southeast Asia where there were fewer palms that needed to be greased. Fortunately, President Joko Widodo, is in the process of implementing a major anti-graft campaign. It remains to be seen, though, how this reform will work in practice.

8.6 Policy Recommendations for Malaysia

For Malaysian sites included in this study, there are major issues implementing top-down governance in the parks. I have four policy recommendations. First, Malaysian Marine Parks must reform the way it recruits park staff and officers. The current system attracts too many people who are indifferent toward or incapable of completing their duties required. Second, there needs to be an urgent and aggressive push to end diver damage to corals under water. This could be modeled on the successful efforts of Sabah Marine Parks where they train and hire underwater officials to prevent damage to corals. Third, predictable and egregious rules violations must come to an end. This includes Marine Parks looking the other way when trawlers invade their waters in the off-season. Fourth and most importantly, there is an urgent need to engage the community. The Malaysian Marine Parks were established without the community's consent. They are fighting a long history of tacit opposition, although short-term programs such as the United Nations Environment Programme's (UNEP's) community engagement program discussed in the earlier chapters are beginning to make headway. There is a need for new forms of community engagement that blend religious and environmental values, as they do in Bali, and employ locals to participate in patrols and conservation work.

8.6.1 Recruiting the best people and increasing scientific expertise

The Malaysian government should reconsider the way it recruits Marine Parks officers. First, the application process should differ from what is required for the majority of civil service jobs because of the dedication and scientific knowledge required for Marine Parks officers. The University of Malaya with its notable marine science and conservation department could be a source of new recruits. Perhaps the government can devise a competitive scheme whereby marine science programs can send their best students upon graduation to work as officers in Marine Parks. As it currently stands, the apathy of Marine Parks employees compared to the passion of many young Malaysian conservation leaders is a gap where there may not need to be one.

8.6.2 Underwater patrols and increased fines

If you find yourself on an island in a Malaysian MPA, you will see the same dark blue posters in every building. These tell you not to touch the wildlife. They are written in every language. Yet, survey respondents felt that there was a serious gap in Marine Parks' ability to educate visitors and local stakeholders. Many respondents in the dive industry recalled countless events when tourists broke rules. Although it is possible to ask them to correct their behavior, the scale involved is too great. In other words, more needs to be done to ensure that visitors internalize how important it is not to touch coral or harass wildlife. I have two recommendations regarding information sharing on the part of Malaysian Marine Parks. First, there is a need for an increase in capacity to enforce the rules that are posted throughout the park. All it takes is one dive in a rogue shop to see rampant touching and wildlife harassment. If there were underwater Marine Parks officers patrolling, issuing fines for rule-breaking and making sure these fines are collected, environmentally damaging behavior would undoubtedly decrease. Some might argue this is too expensive or labor intensive, but Malaysian MPAs have peak visitation periods and key reefs that are jammed with boats between certain daylight hours. Even one underwater patrol per week could reduce the willingness of divemasters to be permissive of damaging behavior with the knowledge that Marine Parks might issue them a fine.

Second, it must be recognized that posters are an ineffective means of communicating behavioral standards to divers. Individual shops must do more to educate their divers about the seriousness of coral damage, kicking sediment, and "distracted diving" where divers bump into coral because they are focused on their cameras. Many dive shops visited for this study lacked accountability for their divers and need to work harder to prevent bad behavior. They should provide additional training to dive guides with limited English skills so that they are able to communicate to clients on how serious coral damage is. It may require a more multimedia approach to educating divers, such as a short video that explains the ecology of corals to those seeking to dive in the Marine Parks. As it stands now, the behavior of divers does not reflect the fact that coral is sensitive and takes hundreds of years to grow into a reef. This understanding needs to be better conveyed to visitors in Malaysian Marine Parks. Such materials must be provided in Chinese given the influx of tourists from China.

8.6.3 Highly visible violations need to be stopped

There are several highly visible and egregious violations of MPA rules that are universally known to occur and universally loathed among stakeholders. The most visible and well known occur in the off-season when illegal fishing vessels enter the MPA and fish. Many cite the impossibility of stopping this because the elites involved have ties with powerful politicians. With the rise of the Bersih movement in Malaysia, however, it would not be hard to imagine a growing demand, especially among Malaysian youth, for an end to environmental corruption. For now, the lawlessness that plagues the MPAs in the off season should become the first priority for Marine Parks aiming to improve biodiversity conservation. There are also new and developing high-tech solutions to ensure that illegal fishing does not occur. These include the use of real-time satellite data. Such technical solutions require skilled and dedicated staff as well as a political system in which the rule of law can be expected to prevail over cronyism in the fishing industry.

8.6.4 Courting communities

The most important obstacle to increasing the effectiveness of Malaysian MPAs is its poor track record of community engagement. Communities that inhabit the islands in the MPAs and house the people who make up a majority of the ecotourism economy need to be actively included in conservation efforts as quickly as possible. Young men and women need to learn to understand the importance of reefs. During my fieldwork, I witnessed several outreach programs and educational sessions for local schools run by Malaysian Marine Parks. This youth-directed effort appeared to have a positive impact on Marine

Parks, but there is still pronounced apathy among many who dwell in the villages within the MPAs. There needs to be more outreach that explicitly links livelihoods and the health of reef ecosystems. Such a program must begin with the young men who run the water taxis. They are highly visible to tourists and work full days on the water. There is no reason why these men, when asked, should have perceptions that reef ecosystems and their livelihoods have no relationship. Additionally, religious leaders can do a better job of linking the Islamic faith to environmental conservation. Many Balinese stakeholders noted the importance of religion as a motivator for environmentally conscious behavior. Given the deeply religious villages in Malaysian MPAs, greater involvement by these communities could change the mindset.

8.7 Conservation of Coastal Biodiversity

In the twenty-first century, sound coastal biodiversity conservation is more important than ever. Coastal systems such as coral reefs, linked systems such as mangroves and seagrass in the tropics or saltmarshes in temperate latitudes protect human settlements from disturbances linked to climate change. These include increased storm surges from weather events with greater intensity. Buffering communities from such impacts enables coastal settlements to exist. They act as a simultaneous pathway to not only climate risk mitigation but also adaptation. When sea levels rise and storms increase in intensity due to elevated sea-surface temperatures, these coastal ecosystems are the first line of defense. Thriving fisheries within these systems feed both locals and visitors and create economic opportunities for locals. How can these environments be sustained in developing countries given higher poverty levels and limited government finances? I conclude with one key policy recommendation regarding conservation of coastal biodiversity. Co-managed institutions for biodiversity conservation can be extremely effective for sustaining environmental, social and economic systems in developing countries.

This book has shown how communities with co-managed coastal ecosystems, where local resource users make decisions and enforce them, have high levels of legitimacy, adaptability and innovation. Co-managed MPAs afford local people greater possibilities for managing complex socioecological systems, allowing social, economic and environmental variables to be considered in any management or policymaking efforts. This is not to say that all co-managed systems look exactly like those described here. Rather, the cases I have described can be instructive for other countries looking to support community-based management. Development agencies, NGOs and governments can and should learn from Balinese example. The way that Balinese MPAs have forged strong rules and expectations, linking livelihoods and environment, may serve as a compelling example for governments seeking to implement their own co-management regimes.

Three key elements of Balinese MPAs appear to explain their effectiveness. These include closing certain jobs within the lucrative diving sector to outsiders. This ensures that the benefits of ecotourism go to local residents, providing jobs across all sectors, education levels and allowing businesses to be active stakeholders in building privately funded projects and linking them to a local "culture of conservation." Everyone who resides in coastal villages should know the importance and value of reefs. In communities with transitional economies such as those where fishing used to dominate, but hospitality and diving are now ascendant, there was a risk that foreigners, especially Westerners, could move in and take all these higher-skilled, higher-paid jobs that required English speakers. In all villages surveyed for this study, there exists some form of protectionism locally to ensure that key jobs are retained for villagers.

Admittedly, rules are broken. Reserving jobs in diving and hospitality, such as the lady porters of Amed where women who speak almost no English and have no formal education are guaranteed a role in the dive sector and a stake in the MPA's success, demonstrates how to combat community apathy toward coral reefs across socioeconomic classes. Teaching people about the importance of reefs using posters needs to be complemented by offering them a financial reason to care about reefs and a guaranteed livelihood linked to vibrant reefs. Finally, in Balinese MPAs, businesses joyfully partake in the culture of conservation, with many hotels financing artificial reefs for community training programs for teenagers in the village who want to learn to dive. These stakeholders are not excluded from participation by a bureaucracy that leaves value on the table and denies them a stake in conservation. Many hotel and dive shop owners double as some of the most active champions for conservation in Indonesia. The largest lesson presented by Indonesian MPAs is that involving a wide range of stakeholders in marine conservation produces many benefits.

Appendix A

RESEARCH DESIGN

A.1 Overview

This research uses institutional theory in order to systematically define the analytical constructs of central and decentralized institutions, marine protected areas (MPAs), along the eight design principles described by Ostrom. It also uses a socioecological systems framework to compare these institutions along specific criteria. Case sites were selected, or sampled, based on *theoretical grounds*, specifically on key "concepts, properties, dimensions, and variations" of sites (Corbin and Strauss 1990, 420). In other words, institutions were defined using the literature, differentiated between types and then different specific MPAs were studied. This research draws its design elements from grounded theorists Corbin and Strauss who argue that a project begins with researchers having an idea of a phenomenon they want to study—in this case top-down and bottom-up MPA institutions—then based on the idea, they select a group of MPA institutions to compare. Once in field sites, researchers are not sampling individuals; instead, they are sampling for "incidents, events or happenings" denoting in the case of this particular study local level strategies for MPA management across stakeholder groups (Corbin and Strauss 1990, 421).

A.1.1 Creating analytical constructs: Institutions

Institutional theory was critical in forming the analytic constructs of centralized and decentralized institutions. When designing my research, I divided it into two parts as defined by grounded theorists: the phenomenon and the interactions (Corbin and Strauss 1996). The phenomenon in this study is the institution, either top-down or bottom-up MPAs, whereas the interactions are described using a socioecological systems framework. Besides the more obvious criteria that federal government manages centralized institutions and local government manages decentralized institutions, the component parts of these phenomena had to be defined so that certain comparisons between Indonesian and Malaysian MPAs were systematic. This research design addresses the criticism commonly levied against co-management that it lacks systematic comparative analyses. Grounded theorists emphasize the importance of comparative analysis, arguing that it guards the researcher against bias; helps in achieving greater precision, whereby the original subjects (top-down and bottom-up MPA institutions) may be further subdivided; and helps in improving consistency of analysis (Corbin and Strauss 1990).

I used Ostrom's eight principles of institutional design in order to firmly establish the analytic constructs of institutions, demonstrated in the table below. One of her criteria, the right to organize for those impacted by the resource, was adjusted for the Malaysian case because centralized management transfers power over the right to organize to central government. In the Malaysian case, I instead opted to use the right for stakeholders to organize outside of the formal MPA, such as the ability for stakeholders to engage with nongovernmental organizations (NGOs) or join the staff of Marine Parks Malaysia. Table A.1 shows the models of top-down and bottom-up MPAs within Indonesia and Malaysia according to institutional theory.

A.1.2 Comparing analytical constructs: The socioecological systems framework

Once I was able to construct a model of Malaysian and Indonesian MPAs according to institutional theory, the next step was to theorize the interactions that these phenomena cause (Corbin and Strauss 1996). I opted to use a socioecological systems framework that has three parts. These are as follows: whether institutions allow for an integrated form of management focused on ecosystems, economies and societies; whether the institution creates opportunities for stakeholder participation, and if it does not, whether it is perceived to have legitimacy; and whether the institution has adaptive capacity whereby previous decisions inform new decisions and inventive solutions to socioecological problems are possible. These three criteria were used to design survey and interview instruments.

For the first part, integrated management, I wrote questions that linked conservation, livelihoods and the management institution in order to measure whether stakeholders perceived management as a socioecological endeavor. For the second part, participation and legitimacy, I wrote questions that asked whether stakeholders could participate. In order to not bias my analysis against top-down management where participation is not allowed, I included questions asking stakeholders whether they perceived institutions as legitimate. I used a modified definition for legitimacy from institutional theory that defines it as a state where resource management decisions are shown to accomplish

Table A.1 Defining constructs: Institutions for resource management.

Principle	Indonesia	Malaysia
Level of government responsible for MPA	Regency governments (local)	Federal government (central)
Clearly defined boundaries: households and individuals allowed to withdraw from a resource and the defined territorial borders of the MPA.	No fishing within two kilometers of the shore. Villages form institutions for reefs that line their shores.	No fishing within two nautical miles of the shore. Marine Parks officials drew contemporary territories in the early 1990s.
Rules: these govern and limit the use of a resource and are must be appropriate for local needs.	No fishing within two kilometers from shore, only locals can hold certain dive industry positions, only locals can own boats that bring tourists to dive and snorkel sites, boatmen take turns bringing tourists to sites. Village institutions make rules for reef visitors that include no trampling or touching corals or organisms, no littering and no taking of shells or corals.	No fishing within two nautical miles of shore. Marine Parks make the rules for accessing the reefs. These include no touching or trampling corals, no fish feeding and no taking of shells or corals.
Collective action: Can stakeholders affected by management and managers come together to change rules if they must?	Every village has its own management institution for its MPA. These institutions engage in collective action.	Only Marine Parks employees can change rules, but they do engage NGOs and university partners in management decisions.
Monitoring: the system in place for conducting surveillance on the resource to ensure rules are followed.	Village MPA institutions are in charge of monitoring.	Marine Parks officers patrol the Malaysian Marine Parks.
Graduated sanctions: the presence of punishments for different levels of rule breaking.	There are graduated sanctions in Indonesian villages for different forms of rule breaking. If boatmen skip others whose turn it was to bring tourists to dive sites, their boat can get impounded. If dynamite fishermen are caught, they can go to jail. If less damaging types of fishing are discovered, those rule breakers may pay a small fine.	Different levels of punishment are in place for locals who break smaller rules such as pole and line fishing; punishments are small and could include fines. For a boat caught trawling in the MPA, the punishment could include jail time.

Continued

Table A.1 *(Continued)*

Principle	Indonesia	Malaysia
Conflict resolution mechanism: resource users and managers can access a low cost and rapid system to resolve conflicts.	Monthly meetings characterize Indonesian MPAs where conflicts can be resolved and problems discuss.	Conflict resolution mechanisms begin with the Marine Parks office, which is conveniently located on each MPA site. Stakeholders can bring issues to the office and have park employees settle them. If issues are too large for Marine Parks offices, they can ask the central office in Kuala Lumpur to assist in resolving conflicts among stakeholders.
Rights to organize: those who depend on the resource can organize institutions for local resource management.	In Indonesian co-management, the right to organize is held by villages. In all three villages studies here, the pre-existing fishermen's cooperatives became the institutions where reef management MPAs were parked, since these men were the main sources of transitioning work forces in the early 1990s, from fishing to tourism.	The right for stakeholders to organize in Malaysia does not exist, but Malaysian Marine Parks do involve academics and NGOs in scientific data collection. NGO activity and environmental activism is not banned or punished in peninsular Malaysia. Resource users can join the Marine Parks as an employee, but they need to take the civil service exam. Literature suggests resource users prefer government to manage the resource.
Nested enterprises: governance resource is organized on multiple layers of government.	MPAs are linked to village governments, which can be involved if problems happen that the MPA institutional members cannot solve. Regency governments and provincial governments are also linked to these lower level institutions, but capacity is thin the higher one goes in Indonesian governance.	Marine Parks are top-down, but they have offices on site within each of their MPAs. In 2004, there was a bureaucratic streamlining that was meant to make it easier for state governments to work collaboratively with their marine park office. As was mentioned before, large problems requiring expertise or financing can be sent to central offices in Kuala Lumpur.

stated objectives, or where management decisions are made in *appropriate ways* (March and Olsen 1989). Legitimacy, or a lack thereof, is a frequent reason why natural resource management institutions receive criticism. Reasons can be that institutions are rigid, incompetent, too powerful, intervene too much or serve special interests (March and Olsen 1989).

In writing interview and survey questions on legitimacy, I chose between two distinct conceptualizations of legitimacy as differentiated in Buchanan and Keohane (2006). A normative definition of legitimacy describes a ruling institution as one that has the "right to rule," secure compliance and ability to issue sanctions for non-compliance. Sociological definitions of legitimacy focus more on the perceptions of stakeholders themselves, asking whether stakeholders "widely believe" that institutions have the right to rule (Buchanan and Keohane 2006, 3). For my work, I adopted the second definition of legitimacy, since most of the people whom I sought to interview and survey were local resource users in the villages I was studying. Legitimate institutions as defined in this research would not necessarily involve all stakeholders in decisions, but instead use navigate predetermined social commitments, predetermined worldviews, and "standard operating procedures" in a host society (Adger 1999). Sociological interpretations of legitimacy can also relate to whether institutions contain "traditional notions of power" where stakeholders see their leaders as authentic sources of authority (Dodson 2014).

For the final part of the socioecological systems framework used to compare MPAs, I examined the adaptive capacity of the institutions. I wrote questions that asked stakeholders to comment on three aspects of adaptive capacity. The first was on the role of learning, or whether they use past interventions in the ecosystem to inform management. The second was on the capacity of the institution to change programs or efforts it sees as not working. The third was on whether institutions support innovative problem-solving to address ecological problems.

I drew my questions on adaptive capacity from the literature on *resilient institutions*. Resilient institutions are social arrangements aimed at increasing the ability of socioeconomic and ecological systems to absorb unanticipated disturbances without flipping into a qualitatively different and undesired state (Gunderson 1999). This capacity to cope with disturbance and change while still retaining critical functions, structures and feedback mechanisms is applicable not only to socioeconomic systems but also to ecological systems (Berkes and Folke 1998; Ingold 2000; Berkes et al. 2003; Olsson et al. 2004; Berkes 2007). Resilience, in an institutional setting, is the ability to adapt and change in order to respond to disturbances, also known as its adaptive capacity (Armitage et al. 2007).

In order for institutions to be resilient to unanticipated disturbances they must adapt, and adaptation requires some degree of reflecting on what has been learned. The ability to learn in an organizational setting is something that has received a lot of scholarly attention (Armitage et al. 2007; Susskind et al. 2012). Organizational learning occurs when organizations acquire new understandings, know-how, techniques and practices. Adaptive capacity requires institutions to be able to learn, be flexible enough to reflect on previous responses to shocks, and be open to adopting novel solutions to challenges (Walker et al 2002).

A.1.3 Measuring ecological output

This dissertation uses institutional characteristics as its independent variables and ecological outputs as its dependent variables. Ecological health of a coral reef can be measured in a variety of ways that vary by cost and time. I opted to use living coral cover as my indicator of ecological health since it fits both local degradation histories and economic patterns while also falling within practical limitations of funding and time. Technical details on these surveys are included in Section 2.4. Percentage of living cover coral is a useful, albeit coarsely grained, indicator for large-scale, high-impact damage, such as dynamite and cyanide fishing that occurred extensively across Southeast Asia until governments began declaring MPAs in the 1980s and 1990s. Coral reef field survey protocols suggest that percentage of cover is both a viable economic *and* ecological indicator of reef health for rapid assessment and monitoring needs (Hill and Wilkinson 2004). This is especially true in places with economies that depend on reef tourism, where visitors expect to see living corals. Large amounts of living coral cover also help reefs deliver maximum amount of ecosystem services to communities who depend on them for recreation and tourism (Hill and Wilkinson 2004).

A.1.4 Problems with research design

There are lingering issues with causality in this design, specifically over alternative explanations for ecological observations. Alternative explanations could include climate change, where my observations may have occurred in years with unseasonably high or low temperatures; outlier weather events such as a year with particularly strong storms that damage reefs in unexpected ways; and differences in larger-scale processes such as shipping lanes further offshore or land-use planning processes that can have large impacts on environmental quality unrelated to management institutions. In order to address these issues, I triangulated my ecological data with the few studies that exist on these sites from the published literature on coral reef ecosystem health and NGO reports on these field sites. Although this does not resolve criticisms on endogeneity altogether, it diminishes the likelihood that what this study observed was the product of a random event. Additionally, this study's design follows similar study designs as described in the Introduction seeking to link ecological outcomes and institutional design.

A second criticism could be that such a small number of cases (five) in this study may not yield useful data for the generation of general policy recommendations. One way to address this criticism is to select cases carefully controlling for as much as possible in order to eliminate alternative explanations. Careful case selection, notes Flyvbjerg, with intense observation have shown repeatedly through history to offer findings that are just as valuable as those of larger n studies designed according to accepted statistical sampling protocols (2006). Flyvbjerg also notes that comparative case studies can test theory as long as theory can be related to a particular interpretation of a case (2006). Thus, both institutional and socioecological systems theories were heavily drawn from in the formation of analytical constructs and interview/survey instruments.

Appendix B
DATA AND METHODS

B.1 Overview

This study uses a mixed-methods approach with both qualitative and quantitative analysis. Mixed-methods approaches are studies with qualitative and quantitative data collected sequentially or concurrently (Hanson et al. 2005). Mixed methods studies are thought to use both types of data in order to enrich their results in ways that only one form of data would not allow (Brewer and Hunter 1989). For example, using a combination of qualitative and quantitative data may allow for generalizable findings with deeper understandings of phenomena of interest and interactions. Additionally, mixed methods allow for testing theory while simultaneously modifying theories based on the feedback of participants (Hanson et al. 2005).

B.2 Interviews

I collected three types of data at each field site. For the first type, I used long-form, semi-structured interviews that asked key informants detailed questions on three topics that include (1) the design principles of their local MPA institution, (2) mapping stakeholder groups defined for convenience in Sections 4.4 and 4.5, and (3) the three-part socioecological systems framework for comparing institutions. The interviews had two purposes: figuring out how informants defined and categorized institutions and stakeholders, and asking informants to respond to questions for which I had already defined terms (Harrel and Bradley 2009). Interview questions on design principles and stakeholders were focused on defining and categorizing, whereas interview questions on socioecological systems interactions were predefined by me and based on the literature. Interviews are a time-consuming method for data collection, yet they provide the most detailed and complete responses to questions with the greatest potential to discuss conflicting or complex topics with stakeholders (Harrel and Bradley 2009). The need for interviews became apparent in my first visit to my field sites in the summer of 2013, where

I observed that although Lovina had less healthy reefs compared to Pemuteran, its institutions seemed more legitimate and developed than those of Pemuteran (Dunning 2015). This type of contradiction made interviews critical.

Qualitative methods theorists argue that interviews are not only where data is collected but also where participants can co-construct knowledge with the researcher. Doucet and Mauthner argue that interviews are where "identities are forged through the telling of stories and meaning-making begins" (2008, 335). Co-creation of identities and narratives were important to this project for two reasons. When searching for stakeholders to survey and interview, I found that stakeholder-generated identity categories of different roles within the MPA were far more accurate than my own and helped in communicating additional needs for respondents to stakeholders who introduced me to peers. When analyzing my statistical data from surveys, I found that stakeholders' narrative stories were important in offering explanations of results as well as necessary context that allowed me to explain results.

I conducted an average of 35 interviews per site focused broadly on local resource user stakeholders and their family members who earn their income from reef tourism. This group included members of the MPA management organizations, resource user stakeholders such as dive industry workers, general tourism workers, village residents, village leaders, government officials across scales, NGO workers, scientists, environmental activists and hotel owners and employees. I conducted 36 interviews in Pemuteran and 37 in Lovina in May through August of 2013, 30 interviews on Tioman Island in August and September of 2014, 33 interviews in Amed in June and July of 2015 and 30 interviews in the Perhentian Islands in August and September of 2015.

Respondents were sampled for interviews in a snowball method, where key informants, beginning with local leaders, were asked to recommend other knowledgeable stakeholders to speak with. Interviews ranged from 40 minutes to four hours, and included four focus groups in Indonesian and three in Malaysian sites. Focus groups, or group discussion of interview topics, arose informally but were quickly included into the project because of the Southeast Asian cultural propensity to engage in more detailed and lively conversations when friends and peers were included. Figure B.1 and Table B.1 depicts the breakdown of stakeholder respondents with the appropriate categorizations developed collaboratively with respondents.

Interviews included the following 22 questions, and 40 percent of the interviews were conducted in Indonesian and 60 percent in English. A translator was occasionally used on an informal basis in Indonesia, where a nearby person or participant would help with translation. Formal translators were used in both Malaysian field sites for all interviews conducted in Malaysian. Interviews were transcribed while they occurred and were not

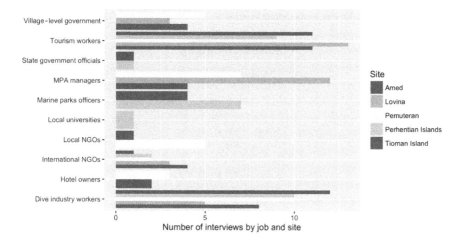

Figure B.1 Interview respondent breakdown.

recorded because of stakeholder discomfort with recording. Respondents were ensured full confidentiality because environmental conservation issues can be somewhat controversial in Indonesia and Malaysia. All respondents were asked to give verbal informed consent to the interview and were instructed to say "pass" when they did not know the answer to the question. In Southeast Asian culture, "saving face" is an important value and stakeholders may not admit when they do not know an answer to a question. Instructions were given to respondents informing them that not knowing an answer was acceptable and to say "pass" when they did not know (Figure B.2).

It is important to point out the difference in questions that I used to interview Malaysian and Indonesian stakeholders under the subtopic of legitimacy. If I were to ask stakeholders in Malaysian MPAs whether there was power-sharing, the answer would be no, and Indonesian MPAs would look more legitimate. Thus, I needed a culturally appropriate definition of power-sharing, which I drew from the literature on legitimacy.

The specific structure of power-sharing can vary across a wide spectrum according to how much participation government affords stakeholders. On the low end of the participation spectrum (Figure B.3), resource users can simply be passive recipients of information distributed by a government agency. Moving across the spectrum, resource users can be consulted for their views but ultimately left out of the decision-making process itself. On the opposite end of the spectrum are co-managed or community-based management institutions (Ostrom 1990; Reed 2008; Pomeroy and Douvere 2008), where stakeholders do make substantial management decisions.

Table B.1 Stakeholder respondent breakdown.

Location	Stakeholder group	Count
	Dive or snorkel workers	5
	International NGOs	3
Lovina	MPA managers	17
	Various tourism workers	8
	Village-level government	3
	Dive industry workers	10
	Hotel owners	3
	Village-level government	5
Pemuteran	International NGOs	2
	Local NGOs	2
	MPA managers	7
	Various tourism workers	4
	Dive industry workers	10
	International NGOs	4
Amed	MPA managers	4
	Various tourism workers	11
	Village-level government	4
	Dive industry workers	10
	International NGOs	2
	Local universities	1
Perhentian Islands	Marine Parks Officers	7
	State government officials	1
	Various tourism industry workers	9
	Dive industry workers	12
	International NGOs	1
	Local NGOs	1
Tioman Island	Local universities	0
	Marine Parks Officers	4
	State government officials	1
	Various tourism industry workers	11

B.3 Surveys

For the second part of data collection, I administered a brief 10-question survey questionnaire focused on socioecological systems interactions within MPAs. I opted to use surveys for three reasons: first, to measure the statistical significance between different stakeholder responses; second, to allow triangulation; and third, to enhance the precision by using uniform definitions for concepts that can be interpreted comparatively between sites. Statistical significance deals with sampling error, or the differences between values calculated from a sample, and those in the population for a sample statistics such as a mean or a proportion. Sampling error is reduced by

Theme 1: Governance mode
1. What level of government manages MPAs? (Clarification in Indonesia: village, province or central. Clarification in Malaysia: village, state or federal.)

Theme 2: Institutional design principles
2. What households are allowed to engage in reef-based tourism, including snorkel and dive tours, operating boats, owning shops, etc.?
3. Who can make rules for how reef tourism workers access the reefs?
4. Can reef tourism stakeholders change the rules for reef management?
5. Who monitors the reef to make sure poaching and illegal fishing do not occur?
6. Are there punishments for violators? (Follow up: Are there different levels of punishment for someone using dynamite versus someone using a spear gun?)
7. Is there a way for reef tourism stakeholders to resolve conflicts and disagreements?
8. *Indonesia only*: Can reef tourism stakeholders participate in management decisions?
Malaysia only: Do you think that Marine Parks is effective at doing its job?
(The different questions are included so as to not bias the analysis against top-down marine parks, which do not invite all resource user stakeholders).
9. *Indonesia Only*: If locals cannot solve a problem, can they involve higher levels of government?
Malaysian Marine Parks officers only: If you have a problem that you cannot solve on your own, does the central office in Kuala Lumpur help solve the problem?

Theme 3: Socioecological systems framework
Sub topic 1: Integrated management
10. Do you see environmental conservation and your livelihood as being linked?
11. Do reef tourism businesses invest in projects that help conservation?
12. Does the reef management organization respond to crises (such as illegal fishing) in a way that helps you continue to earn a living based on reef tourism?
13. Do NGOs assist MPA managers or business owners in management?

Sub topic 2: Legitimacy
14. Does the reef management organization actually protect the reef?
15. Is the reef management institution something that you value and respect or does it not work?
16. *Indonesia only*: Are there official power-sharing arrangements that devolve responsibility to the village?
Malaysia only: Do managers provide information to visitors and stakeholders for best practices for conservation?

Sub topic 3: Adaptive capacity
17. Something bad happens on the reef (such as anchor damage to corals that requires the institution to act). Do members of the institution reflect on past management interventions before deciding on new ones?
18. The reef management institution sees that one of its programs is not working (for example, it learns that large amounts of poaching happen during a certain time of year). Can it change its program to address this problem?
19. Does the reef management institution support businesses or stakeholders who want to intervene for conservation in innovative or creative ways?

Theme 4: Reef health
20. Is the reef healthier today than in the past? Why or why not?
21. What are some good things that stakeholders do for conservation?
22. What are the challenges that locals face for conservation.

Figure B.2 Interview manual.

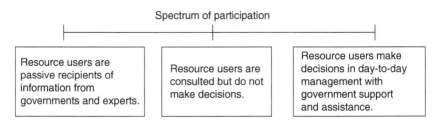

Figure B.3 Spectrum of stakeholder participation from Reed (2008).

increasing the sample size and the *alpha level*, or the level set for the likelihood of statistical significance. I opted for a standard alpha in the social sciences of 0.05 (Lipsky and Hurley 2009). Triangulation is the use of different methods, here qualitative and quantitative methods, in order to study the same phenomena. Researchers can "improve the accuracy of their judgments by collecting different types of data on the same phenomena" (Jick 1979). Surveys are known to be high in reliability, or replicability. This is because they use consistent definitions for phenomena and interactions. High reliability allows for survey data to be used in comparative analyses (King et al. 1994).

I administered oral surveys because one-on-one contact allowed me to explain answers or concepts not understood by the interviewee and to provide a full explanation of the work I was doing. I administered surveys to 54 respondents in Lovina, 53 respondents in Pemuteran, 51 respondents in Amed, 55 respondents in Tioman Island and 51 respondents in the Perhentian Islands. In Indonesia, I conducted these surveys alone, with nearly 80 percent done in English and 20 percent done in Indonesian with no translator, as spoken English is widespread in these two villages. In Malaysia, enumerators were hired in Kampung Pulau Perhentian and Tekek Village. Enumerators were trained in asking for consent and instructing the respondents on how to say they did not know an answer or want to provide an answer. Enumerators were compensated and paid for each completed survey.

Stakeholders were surveyed randomly, in areas where they congregated such as cafes, workplaces or the main stretch of shoreline. This random method generated a number from one to five (n) and surveyed the nth person sitting along the benches and tables in these locations; it also used boat numbers and surveyed randomly chosen boat captains by their boat number. A copy of the short questionnaire is listed in Table B.2. For each question, also included is the hypothesized relationship that the variable has to reef ecological health. Possible answers could include only yes or no, because in pilots, the "do not know" or "neutral" responses were confusing to stakeholders. Instead of including a third option between yes and no, I opted to urge respondents

Table B.2 Predictor variables and their hypothesized effects on reef quality.

Variable	Description	Hypothesized relationship to coral cover
Integrated Management	1. Do you see environmental conservation and your livelihood as being linked?	Communities that perceive links between conservation and livelihoods have healthier reefs.
	2. Do reef tourism businesses invest in projects that help conservation?	Communities where businesses invest in conservation have healthier reefs.
	3. Does the reef management organization respond to crises (such as illegal fishing) in a way that helps you continue to earn a living based on reef tourism?	Sites with stakeholder perceptions that MPAs respond to crises will have healthier reefs.
	4. Do NGOs assist MPA managers or business owners in management?	Sites where NGOs assist MPA managers and business owners will have healthier reefs.
Legitimacy	5. Does the reef management organization actually protect the reef?	In sites where stakeholders perceive the MPA as protecting the reef, there will be healthier reefs.
	6. Is the reef management institution something that you value and respect or does it not work?	In sites where stakeholders value the MPA, there will be healthier reefs.
	7. *Indonesia only:* Are there official power-sharing arrangements that devolve responsibility to the village? *Malaysia only:* Do managers provide information to visitors and stakeholders for best practices for conservation?	In sites with even minimal participation, there will be healthier reefs.
Adaptive problem solving	8. Something bad happens on the reef (such as anchor damage to corals that requires the institution to act). Do members of the institution reflect on past management interventions before deciding on new ones?	In sites where learning is used in decision-making, healthier reefs result.
	9. The reef management institution sees that one of its programs is not working (For example, it learns that large amounts of poaching happen during a certain time of year) Can it change its program to address this problem?	In sites where changes to management are possible, there will be healthier reefs.
	10. Does the reef management institution support businesses or stakeholders who want to intervene for conservation in innovative or creative ways?	In sites where the MPA supports businesses or stakeholders who want to intervene for conservation, there will be healthier reefs.

Table B.3 Summary of surveys and interviews.

Site	Number of interviews	Number of surveys	Dates
Pemuteran	36	54	June 2013
Lovina	37	53	July–August 2013
Tioman	30	51	August–September 2014
Amed	33	55	June–July 2015
Perhentians	30	51	August–September 2015

to say, "pass" or "skip" if they did not know the answer or had no opinions. Stakeholders opted to pass on questions rarely if at all. Table B.3 summarizes the surveys, interviews, places and dates.

B.4 Reef Surveys

In order to measure living coral cover, I combined protocols from several reef survey methods including the *timed swim survey method* used by the Nature Conservancy and the *video manta tow method* described by The Australian Institute for Marine Science (Hill and Wilkinson 2004). This hybrid survey relied on a 20 minute timed swim along a predetermined depth contour (1–4 m at the reef crest and then 5–10 m on the reef slope in 2–3 replicates per depth range) (Hill and Wilkinson 2004, 50). Timed swims took place over stretches of reef that are commonly used by tourists for recreational purposes, chosen by popularity based on promotional materials for the field sites. For some sites (Tioman, Amed and the Perhentians) I was able to take video of the entire timed swim; for other sites (Lovina and Pemuteran) I was able to take either photos or handwritten notes on waterproof paper of cover estimations within a 1 m by 1 m quadrat placed every 30 seconds, with the timer stopped to record coral cover. When video was used, video quadrats were used with filming occurring consistently one meter above the reef.

The timed swim method is a coarsely grained measurement strategy that is regarded as best method to gain a broad understanding and general description of a reef site (Hill and Wilkinson 2004). Percentage of cover was estimated after the dives by viewing the film taken during the swim. Alternative forms of ecological data collection could provide more definitive pictures of reef health, but the time, finances and personnel were not available at this time. Although this survey design was planned for a single diver, a guide was present at each site, which allowed the researcher to focus on filming the coral cover. In Amed and in Tioman, I had assistants helping me perform the survey by monitoring the timing devices, depth gauges or helping me record estimations of coral.

The quantitative output is percentage of cover of living coral, hard and soft, ascertained using visual methods outlined in English et al. (1997, 19). Beginning with the Indonesian sites, in Lovina, I conducted 18 surveys of the main reef in front of Anturan Village used by snorkel and dive tour operators; in Pemuteran, I conducted 20 surveys of three different reefs, Napoleon (seven replications), Temple Garden (six replications) and Close Encounters (eight replications). In Amed, I conducted 23 total surveys of three reefs including Japanese Wreck (five replications), Coral Garden (nine replications) and Tulamben (eight replications). For the Malaysian sites, I conducted 17 surveys on Tioman Island across four reefs including Renggis North (four replications), Chebeh (four replications), Tekek house reef (five replications) and Labas (five replications). In the Perhentian Islands, I conducted 18 surveys

Table B.4 Gomez and Yap (1988) classifications of percentage of cover.

Bad	Medium	Good	Excellent
0–24.9%	25–49.9%	50–74.9%	75–100%

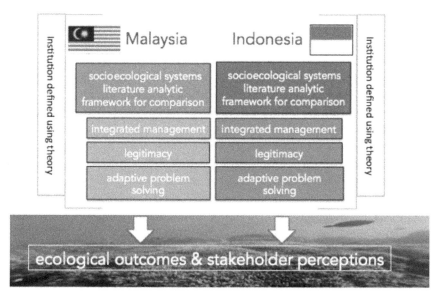

Figure B.4 Research design.

across four sites including Achin Beach (six replications), Tokong Laut (three replications), Shark Point (six replications) and Batu Layar (replications). After the 20-minute timed swim, the video was frozen at the 0:00 mark and the 0:30 mark every minute, resulting in 40 data points per survey that were averaged across replications. The English et al. (1997) visual guide was used to estimate percentage of cover. For sites where data were collected using a quadrat, the timer was stopped manually and measurements taken under water by hand and recorded on waterproof paper.

The measurements were converted into a four-part output of either bad, medium, good, or excellent as defined by categorical thresholds from Gomez and Yap (1988). These categories are defined as in Table B.4.

Figure B.4 summarizes the research design, with its analytical constructs derived from institutional theory, its use of socioecological systems literature as a framework for comparing top-down and bottom up MPAs and the dependent variable as both social and ecological outcomes.

B.5 Analysis

I began by analyzing survey data across field sites. This entailed a two-tailed t-test for statistical significance between two proportions. Samples were independent and gained using random sampling techniques. I assumed unequal variances since this information was not known. The null hypothesis for each survey question was that there was no difference between the proportion of stakeholders answering yes to a question, whereas the research hypothesis said that there was a difference between the proportion of stakeholders answering yes to a question.

For the analysis of interview data, I relied on thematic coding defined as "recognizing and reporting patterns (themes) within data by [organizing] data in rich detail" (Braun and Clarke 2006, 6). A theme "captures something important about the data in relation to the research question" and represents patterned responses of key informants in interviews (Braun and Clarke 2006, 10). The pattern can be observed across interviews given in particular locations, Indonesia and Malaysia, but also across the data set as whole. Finding and describing themes in thematic coding does not mean identifying themes or patterns that exist in certain quantities of data; instead themes are important if they "capture something in relation to the overall research question" (Braun and Clarke 2006, 10), regardless of how much they appear in data.

I specifically applied what is known as *deductive* or *theoretical thematic analysis* (Boyatzis 1998; Hayes 1997; Braun and Clarke 2006). This form of coding is driven by my own theoretical interest in socioecological systems. This form of

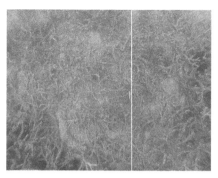

Figure B.5 Ecological data sample—stills from reef survey that show how percent cover was estimated. Photos of Lovina (left) and Pemuteran (right) show representative coral cover. Dark gray shows living coral cover.

coding makes for less in-depth description of all of the data and, instead, drives a more detailed analysis of some components of the data, using theory in order to pare down what is discussed. This type of coding can take two forms, a theoretical approach where you code for the specific research question or the inductive approach where you can allow the research question to somewhat evolve. Because of the appearance of the ecological findings at first glance, I opted for the inductive approach. This meant a focus less on the differences between centralized and decentralized institutions and their ecological outputs and more on stakeholder perceptions (Braun and Clarke 2006, 12).

For analysis of the ecological survey data, images were analyzed according to the percentage of the substrate on which living coral was found to be present. In order to increase accuracy in images that were difficult to judge, a random number generator was created to determine a pixel point on the image where a line was placed (see Figure B.5). Where living coral was present, a red line was marked on the left, and then (when necessary) stacked end to end on the right to better enable an estimate of percent cover. This technique was based on the dot grid method for analyzing photo quadrats discussed in Rogers et al. (1994). Figure B.5 shows this method in use. Bleached coral was counted as living coral, since it is living, however stressed.

Appendix C

CORAL COVER RESULTS

This appendix provides an in-depth look at living coral cover at each site.

C.1 Comparing Coral Cover across Malaysia and Indonesia

In terms of the percentage of living coral cover, the mean coral cover across the two Malaysian sites including the Perhentian Islands and Tioman Island was 44 percent (SD = 0.31) whereas the mean coral cover across the three Balinese sites including Lovina, Pemuteran and Amed was 37 percent (SD = 0.31). Both estimates fall within the "medium" range (on the Likert scale of bad, medium, good and excellent; from Gomez and Yap 1988). Even though this study shows the Malaysian sites as having slightly more coral cover, statistically these sites are not significantly different at the $p = 0.05$ alpha level that would allow us to reject the null hypothesis that there is no difference.

Other studies have shown that living coral cover is not the most responsive ecological indicator when it comes to comparing across management typologies, and this study agrees with those findings (McClanahan et al. 2006). Given the Indonesian sites see 27 percent more of the impact of visitors that Malaysian reefs see, you would expect Malaysian reefs to exhibit coral cover on a different categorical level from Indonesia. Tables C.1 and C.2 display each field site broken down by location, as well as the mean percentage of living coral cover based on an average of 40 individual point estimates per replication, further broken down by individual replications of which there are at least three per site, and summarized with a mean estimate per reef site.

The individual site averages for Malaysia are as follows: 56 percent cover for Tioman Island (SD = 0.14) and 31 percent (SD = 0.28) for the Perhentian Islands. For Indonesia, Amed has 43 percent cover (SD = 0.25), Pemuteran 44 percent cover (SD = 0.17) and Lovina 16 percent cover (SD = 0.11), the only site to dip into the poor category.

Table C.1 Percentage of living coral cover averages by site (Malaysia). Mean % cover is from survey (n = 40).

Reef name	Site name (replication number)	Average % cover	Depth range	Reef average living coral cover
Renggis North	Tioman Island (1/4)	0.75	1–4 m	
2.509308° lat	Tioman Island (2/4)	0.80	1–4 m	M = 0.79
104.08180° lon	Tioman Island (3/4)	0.90	5–10 m	SD = 0.13
	Tioman Island (4/4)	0.70	5–10 m	
Chebeh	Tioman Island (1/4)	0.40	1–4 m	
2.552887° lat	Tioman Island (2/4)	0.50	1–4 m	M = 0.46
104.065304° lon	Tioman Island (3/4)	0.50	5–10 m	SD = 0.22
	Tioman Island (4/4)	0.45	5–10 m	
Tekek House Reef	Tioman Island (1/4)	0.50	1–4 m	
2.462581° lat	Tioman Island (2/4)	0.50	1–4 m	M = 0.45
104.116798° lon	Tioman Island (3/4)	0.40	1–4 m	SD = 0.22
	Tioman Island (4/4)	0.40	1–4 m	
Labas	Tioman Island (1/5)	0.60	1–4 m	
2.901809° lat	Tioman Island (2/5)	0.55	1–4 m	M = 0.55
104.108577° lon	Tioman Island (3/5)	0.50	5–10 m	SD = 0.25
	Tioman Island (4/5)	0.60	5–10 m	
	Tioman Island (5/5)	0.50	5–10 m	
Acin Beach	Perhentian Islands (1/6)	0.50	1–4 m	
5.923482° lat	Perhentian Islands (2/6)	0.50	1–4 m	
102.716781° lon	Perhentian Islands (3/6)	0.00	1–4 m	M = 0.43
	Perhentian Islands (4/6)	0.60	1–4 m	SD = 0.29
	Perhentian Islands (5/6)	0.50	1–4 m	
	Perhentian Islands (6/6)	0.50	1–4 m	
Tokong Laut	Perhentian Islands (1/3)	0.15	5–10 m	M = 0.05
5.944225° lat	Perhentian Islands (2/3)	0.00	5–10 m	SD = 0.11
102.666369° lon	Perhentian Islands (3/3)	0.00	5–10 m	
Shark Point	Perhentian Islands (1/6)	0.30	1–4 m	
5.885880° lat	Perhentian Islands (2/6)	0.70	1–4 m	
102.747109° lon	Perhentian Islands (3/6)	0.50	1–4 m	M = 0.30
	Perhentian Islands (4/6)	0.20	5–10 m	SD = 0.30
	Perhentian Islands (5/6)	0.10	5–10 m	
	Perhentian Islands (6/6)	0.00	5–10 m	
Batu Layar	Perhentian Islands (1/4)	0.80	1–4 m	
5.913298° lat	Perhentian Islands (2/4)	0.80	1–4 m	M = 0.48
102.750444° lon	Perhentian Islands (3/4)	0.30	5–10 m	SD = 0.36
	Perhentian Islands (4/4)	0.00	5–10 m	

C.1.1 Coral bleaching

Every site surveyed in this research had experienced recent upticks in bleaching events. Bleaching, literally the whitening of coral, was first described in the mid-1980s and is correlated with high sea-surface temperatures. Long-term data suggest that bleaching is caused by the gradual increase in ocean temperatures. The bleaching process entails the coral organism ejecting its symbiotic algae, *zooxanthelle*, a significant source of food (Brown 1997). Glynn (1993) states that based on published records of coral bleaching going back to the 1870s, contemporary bleaching episodes are "unprecedented." He distinguishes between short-term bleaching, caused by a variety of stressors such as sedimentation, temperature or light availability, and the large-scale bleaching events that are more indicative of serious global change, primarily climate change (Glynn 1993). Bleaching is known to be a significant driver in the diminishing of coral cover in the past 40 years (Bellwood et al. 2004). There have been two major bleaching outbreaks, 1998 and 2000, which nearly all stakeholders mentioned in interviews (Table C.3). Bleaching and diminished coral cover have a negative impact on recreational ecosystem services of reefs, because without corals, few divers will be interested in spending money to travel to Southeast Asia.

Best practices for reducing the impacts of bleaching are unfortunately large-scale and therefore difficult to achieve. Global climate action in the form of reduced greenhouse gas emissions and a cap to temperature rises is necessary. Local-scale ways that may have an impact on enhancing resilience of reefs to bleaching and speeding recovery time include managing for functional groups described by Bellwood et al. (2004). A functional group is a group of species that perform the same biological function regardless of their taxonomy. In other words, "guilds of fish" across the taxonomic tree can perform the same roles in ecosystem processes. Take the ecosystem process of grazing algae off the surface of coral so that its symbiotic algae can undergo photosynthesis and create food for the coral organism (Bellwood et al. 2004). Another ecosystem process is removing dead corals to allow for the settlement of new, living corals. Functional groups of fish known as scrapers, grazers or herbivores perform these ecological process.

Management decisions can focus on promoting "functional redundancy" on a coral ecosystem, or a condition where multiple species that perform a function are protected as a form of "insurance." If a disease were to devastate one member of the functional group, there are others to fill the role, and the reef will not slip into an undesired state. With bleaching in particular, managing for functional groups is especially important because herbivores, for example, can facilitate the regeneration of reefs after bleaching episodes (Bythell et al. 2000; Bellwood et al. 2004). Specifically, they prepare the substrate for new living

Table C.2 Percentage of living coral cover averages by site (Indonesia). Mean % cover is from survey (n = 40).

Reef name	Site name (replication number)	Average % cover	Depth range	Reef average living coral cover
Japanese Shipwreck −8.34500° lat 115.67890° lon	Amed (1/5)	0.40	1–4 m	
	Amed (2/5)	0.30	1–4 m	M = 0.32
	Amed (3/5)	0.30	1–4 m	SD = 0.19
	Amed (4/5)	0.30	5–10 m	
	Amed (5/5)	0.30	5–10 m	
Coral Garden −8.33730° lat 115.66140° lon	Amed (1/10)	0.90	1–4 m	
	Amed (2/10)	0.80	1–4 m	
	Amed (3/10)	0.80	1–4 m	
	Amed (4/10)	0.80	1–4 m	
	Amed (5/10)	0.80	1–4 m	M = 0.73
	Amed (6/10)	0.70	5–10 m	SD = 0.23
	Amed (7/10)	0.80	5–10 m	
	Amed (8/10)	0.60	5–10 m	
	Amed (9/10)	0.60	5–10 m	
	Amed (10/10)	0.50	5–10 m	
Tulamben −8.27830° lat 115.59720° lon	Amed (1/9)	0.30	1–4 m	
	Amed (2/9)	0.30	1–4 m	
	Amed (3/9)	0.30	1–4 m	
	Amed (4/9)	0.30	1–4 m	M = 0.24
	Amed (5/9)	0.20	1–4 m	SD = 0.22
	Amed (6/9)	0.20	5–10 m	
	Amed (7/9)	0.20	5–10 m	
	Amed (8/9)	0.20	5–10 m	
	Amed (9/9)	0.20	5–10 m	
Napoleon −8.123755° lat 114.664253° lon	Pemuteran (1/4)	0.50	1–4 m	
	Pemuteran (1/4)	0.50	1–4 m	
	Pemuteran (1/4)	0.50	1–4 m	SD = 0.28
	Pemuteran (1/4)	0.70	1–4 m	
Temple Garden −8.134320° lat 114.638964° lon	Pemuteran (1/6)	0.30	1–4 m	
	Pemuteran (2/6)	0.35	1–4 m	
	Pemuteran (3/6)	0.30	1–4 m	M = 0.31
	Pemuteran (4/6)	0.20	1–4 m	SD = 0.19
	Pemuteran (5/6)	0.30	1–4 m	
	Pemuteran (6/6)	0.40	1–4 m	
Close Encounters −8.118313° lat 114.655823° lon	Pemuteran (1/7)	0.40	5–10 m	
	Pemuteran (2/7)	0.40	5–10 m	
	Pemuteran (3/7)	0.45	5–10 m	M = 0.46

Continued

Table C.2 *(Continued)*

Reef name	Site name (replication number)	Average % cover	Depth range	Reef average living coral cover
	Pemuteran (4/7)	0.50	1–4 m	SD = 0.32
	Pemuteran (5/7)	0.50	1–4 m	
	Pemuteran (6/7)	0.50	1–4 m	
	Pemuteran (7/7)	0.50	1–4 m	
House Reef	Lovina (1/19)	0.20	1–4 m	
−8.148653° lat	Lovina (2/19)	0.20	1–4 m	
115.034439° lon	Lovina (3/19)	0.20	1–4 m	
	Lovina (4/19)	0.20	1–4 m	
	Lovina (5/19)	0.10	5–10 m	
	Lovina (6/19)	0.10	5–10 m	
	Lovina (7/19)	0.10	5–10 m	
	Lovina (8/19)	0.10	5–10 m	
	Lovina (9/19)	0.10	5–10 m	
	Lovina (10/19)	0.10	5–10 m	M = 0.16
	Lovina (11/19)	0.10	1–4 m	SD = 0.18
	Lovina (12/19)	0.10	1–4 m	
	Lovina (13/19)	0.20	1–4 m	
	Lovina (14/19)	0.20	1–4 m	
	Lovina (15/19)	0.30	1–4 m	
	Lovina (16/19)	0.40	5–10 m	
	Lovina (17/19)	0.40	5–10 m	
	Lovina (18/19)	0	5–10 m	
	Lovina (19/19)	0	5–10 m	

polyps to colonize, the first step to the reef's regeneration (Bythell et al. 2000). Additionally, management balances the functional groups, where, for example, rules are made to ensure that not too many predators (such as triggerfish) are removed in order to keep eroder populations (such as sea urchins) low to give juvenile corals a chance to re-grow (McClanahan 2000). Management focused on functional roles requires some degree of scientific expertise. The literature on functional group-oriented management strategies still remains in the theoretical changes, however.

C.1.2 Ecological impacts of the dive industry

I have made the argument that given that Indonesian sites receive more visitors, their reefs face additional stress. I have included in this section photos from

Table C.3 Bleaching episodes in Indonesia and Malaysia.

Year	Areas impacted and a brief description
1997–1998 (El Niño)	This first major bleaching event reduced to overall health of Indonesian reefs (Burke et al. 2002). In Indonesia, this event impacted Sumatra, Bali, Java, Lombok and Timor after beginning in Papua. Many central Indonesian reefs were spared because of high upwelling (Goreau et al. 2000). An estimated 16% of global coral was killed. This was the largest bleaching episode in Bali, where 50% of Balinese coral was bleached overall and 100% of coral in Bali Barat National Park (ICRI 2010). Habibi et al. (2007) say that 5–19% of Indonesian corals bleached in this event. Goreau et al. (2000) say bleaching rates in Bali were at 50% but lower in islands further east thanks to upwelling. Peninsular Malaysia did not experience this event, but East Malaysia did (Goreau et al. 2000). Bleaching during this event in Malaysia was confined on isolated patches (Wilkinson 1998; Tan and Heron 2011). Nearby Singaporean reefs did experience this event (Goreau et al. 2000).
2009–2010 (El Niño)	Indonesia, Malaysia, Thailand and the Philippines saw large-scale bleaching. Specifically, Indonesia, Java, Bali, Lombok, Sulawesi and West Timor were impacted. Acropora corals were the most heavily impacted, with 80% mortaility in some sites (ICRI 2010). In the field sites included in this study, Reef Check Foundation Indonesia found Amed to have 40% bleached corals, with 10% in Amed (Habibi et al. 2007). The Wildlife Conservation Society (WCS) found 60% bleached coral in Sumatra. Malaysia also saw bleaching in the 2010 event, with specific sites in this study closed by Marine Parks because of the intensity of bleaching (Laman Web 2016). Closed sites included Tioman's Chebeh and Batu Malang (with 60-70% and 70-80% bleaching respectively), and various reefs on the Perhentian Islands which bleached at the rate of 70% (Laman Web 2010). Over all of peninsular Malaysia, 12 of 83 dive sites were closed (Rakyat Post 2014). Tan and Heron found that 50% of corals in Tioman and the Perhentians were bleaches, with two-thirds of those completely white. They suggest the scale of bleaching related damages is increasing (2011).
2015–2016 (El Niño)	The US National Oceanic and Atmospheric Administration (NOAA) announced the third global bleaching event this year. The media has covered bleaching in Hawaii, Florida, the Caribbean and the Australian Great Barrier Reef. It has yet to be seen whether this event will impact Indonesia and Malaysia. NOAA predicts bleaching will impact Indonesia (Kennedy 2015).

various field sites showing some common diver behaviors across Indonesian and Malaysian sites that negatively impact reef health. Often, when dive tourism is compared to economies that rely on fishing, it is framed as the more ecologically sound choice. I saw dozens of instances of divers kicking sediment onto living coral, touching coral, breaking coral with their tanks and

Figure C.1 Negative dive industry impacts. Clockwise from top left: (1) Diver kicking sand onto living corals (in basic training divers are instructed not to do this so as to not smother corals). (2) Divemaster and clients touching a massive coral. (3) Divemaster standing on a coral. (4) Diver kicking plating coral. (5) Diver, who could not swim unable to ascend for her safety stop so, she stayed on the bottom and held onto corals until the divemaster rescued her.

trampling coral (Figure C.1.1). This included the occasional divemaster or dive shop owner.

C.2 Comparing Coral Cover Findings with Other Studies

Several studies were referenced in order to make sure that other research supports the findings in this dissertation. Beginning with Indonesia, the most comprehensive survey of Balinese reefs is the *Bali Marine Rapid Assessment Program*[1] (Mustika 2011). This survey found medium coral cover in its sites in Pemuteran, as well as medium coral cover in its sites in Amed and Tulamben. These findings are similar to the coral cover findings in this dissertation. In addition, the United Nations Environment Programme (UNEP) case study on sustainable fisheries examines coral cover in several of the same field sites as this study (2005). It also finds medium coral cover in Pemuteran (23–38 percent) and poor coral cover in Lovina on the main reef of Anturan (7–20 percent). I found 31–55 percent coral cover in Pemuteran and 16 percent cover in Lovina, which are compatible findings.

There are also studies of living coral cover on several of the Malaysian sites examined in this dissertation. Yewdall et al. (2012) reported an average of 40–45 percent living coral cover on Tioman Island, placing it in the medium range, findings that are supported by the results of this study. They found 25 percent cover on Labas, 84 percent cover on Renggis North and 21 percent cover on Chebeh, or an average of 43 percent. I found 55 percent cover on Labas, 79 percent cover on Renggis North and 46 percent cover on Chebeh, or an average of 60 percent. Although my three percentages are somewhat higher, overall averages across the site as a whole are similar. Additionally, Reef Check Malaysia provided me with their 2014 Tioman Island survey data that showed 60 percent cover on Labas, 80 percent cover on Renggis North and 49 percent cover on Chebeh, or an average of 63 percent which is very close to my own observations of these sites. Reef Check Malaysia found 49 percent cover at Tekek house reef and I found 45 percent. Reef Check Malaysia's 2010 *Annual Survey Report*[2] places all of Malaysia's reefs in the medium category

---

1 Mustika (2011) used a point intersept transect in order to collect their data, with two 50 m transects placed side by side at two different depths. Benthic surveys were performed at 50 cm intervals.

2 It is important to note that I used a timed swim and either video quadrats or a quadrat when no video option was available, whereas the Reef Check Malaysia survey method is a 100 m transect along two depth contours. They collected point intercept data at four points of all substrate type, including living soft and hard corals. Likewise, other studies used a version of the point intercept method employed by Reef Check.

of percent cover at 44.31 percent. It places the mean percentage living coral cover on the Perhentian Islands at 35.78 percent and that of Tioman at 60.47 percent. These findings are also quite close to my own. Toda et al. (2007) found an average of 38.55 percent coral cover on the Perhentian Islands, and 44.77 percent cover on Tioman Island, findings that also fall into the medium level category and match the findings of this study. Table C.4 shows all of the findings of these similar studies next to my own findings in a consistent format. Some of the information is repeated for ease of comparison.

In addition to living coral cover, other studies on my field sites include data on alternative ecological variables that can be used to give a more detailed picture of reef health, and triangulate my findings. For example, in Reef Check Malaysia's benthic surveys of coral reef substrate, they found Malaysian sites to be in the "fair" category. A large amount of rock, much of which is dead coral, is the result of the 2010 bleaching event that severely impacted Malaysian Reefs (Reef Check Malaysia 2010). Additionally, they publish data on nutrient indicator algae, which would signal that the poor state of sanitation in Malaysian Marine Parks that I describe in later sections is impacting reefs negatively. Nutrient indicator algae is surprisingly low, given how many informed stakeholders were concerned about the neglected state of the septic tanks that characterized many tourist facilities on the Malaysian MPAs (Reef Check Malaysia 2010). One last set of data on fish population abundance suggests that fish populations within important target groups are in the low range (Reef Check Malaysia 2010). Fish with extremely high values on live fish markets, such as the humphead wrasse, and others such as Barramundi cod and sweetlips were in low numbers and were rarely sighted. Reef Check Malaysia suggests this is due to poor enforcement on targeted fishing in the MPA. In my 35 dive surveys in Malaysian Marine Parks, I only saw three sweetlips. My own informal counts of parrotfish over 15 cm on 10 dives in Malaysian MPAs suggested similar findings, with an average of 2.5 fish spotted per dive. Parrotfish are herbivores and help perform the ecological process of grazing corals. They are a key member of the functional group for post-bleaching recovery. Interviews and observations suggested locals illegally fish parrotfish for food.

Similar to Reef Check Malaysia's report, the Bali Rapid Marine Assessment also provides some additional data that I used to triangulate my findings. For example, it also surveyed coral reef substrate finding levels of rubble that place reef health at "fair" levels, citing most coral mortality to crown of thorns starfish, snail predation and algae growth from excess nutrients in the water from poor sanitation in coastal settlements (2011). Lesser causes of damage included white band disease, fishing damage and seaweed farms. Like Malaysia, the Bali Rapid Marine Assessment also found signs of overfishing

Table C.4 My findings compared to those of similar studies.

My findings	Findings of other studies
Indonesia	
Overall 37.46% coral cover (medium according to Gomez and Yap [1988])	**Mustika (2011)**: Medium coral cover in Amed, Tulamben and Pemuteran according to Gomez and Yap (1988)
Pemuteran: 31–55% cover. Lovina: 16% cover	**UNEP (2005)**: Pemuteran: Medium coral cover (23–38%). Lovina: Poor coral cover (7–20%)
Malaysia	
Malaysia: 44.47% coral cover (medium according to Gomez and Yap [1988])	**Yewdall et al. (2013)**: Medium coral cover on Tioman Island according to Gomez and Yap 1988, or 40–45%
Labas: 55% cover	Labas: 25% cover
Renggis North: 79% cover	Renggis North: 84% cover
Chebeh: 46% cover	Chebeh: 21% cover
Overall: 60% cover	Overall: 43%
	Reef Check Malaysia (2014) Tioman Survey Data
Labas: 55%	Labas: 60%
Renggis North: 79%	Renggis North: 80%
Chebeh: 46%	Chebeh: 49%
Tekek house reef: 45%	Tekek house reef: 49%
Overall: 60%	Overall: 63% (SD = 0.21)
	Reef Check Malaysia's 2010 *Annual Survey Report*
Malaysia overall: 44.47%	Malaysia overall: 44.31%
Perhentian Islands: 31.50%	Perhentian Islands: 35.78%
Tioman Island: 56.25%	Tioman Island: 60.47%
	Toda et al. (2007)
Perhentian Islands: 31.50%	Perhentian Islands: 38.55%
Tioman Island: 56.25%	Tioman Island: 44.77%
	The 95% confidence interval given by this paper is 17–38% living coral cover in Tioman and the Perhentians.

at all sites, with the most valuable species absent altogether. "In over 350 man hours of diving the survey team only recorded [...] three Napoleon wrasse." (Mustika 2011, 4). This suggests that no-take rules for the most valuable fish are not being enforced.

C.3 Images from Surveys and Stakeholder Perceptions on Reef Health

This section provides several images from each survey site to give the reader a sense of what reefs looked like across Indonesian and Malaysian field sites.

C.3.1 Lovina, Indonesia

The images in Figure C.2 are from the main reef in front of Anturan village. The overall average in Lovina for percentage of cover is 16 percent, which falls

Figure C.2 Survey images from Lovina, Indonesia.

in the bad category. Respondents suggested that Lovina's reefs are returning but will take a long time because of a combination of years of destructive fishing, several back-to-back intense storm seasons in the past 10 years and annual bleaching events that, according to many respondents, are growing worse. Nearly all stakeholders mentioned anxieties about bleaching in their interviews. "They come back little by little," said one local MPA leader, whose views were widely echoed in interviews, "but the bleaching gets worse and worse."

C.3.2 Pemuteran, Indonesia

The images in Figure C.3 are from survey sites on Napoleon Reef Close Encounters Reef, and Temple Garden. The overall average in Lovina for percentage of cover is 44 percent, which falls in the medium category. Respondents suggested that Pemuteran's reefs are returning to their previous pristine state and that they look much better than those of nearby Lovina because early tourism entrepreneurs limited development and financed conservation more

Figure C.3 Survey images from Pemuteran, Indonesia. Top left: Pemuteran near shore reef. Top right: Survey on Napoleon Reef. Bottom left: Survey on Close Encounters Reef. Bottom right: Survey on Temple Garden Reef.

than stakeholders in Lovina. "We grew slow here from the beginning. We knew what we had early on because our reefs were very big," said one local hotel owner. Another respondent said, "The reefs get better each year, but I worry there is only so much improvement we can see, because the bleaching gets worse and worse." Many other respondents noted that reefs, while improving, were suffering from climate change.

C.3.3 Amed, Indonesia

The images in Figure C.4 are from survey sites on Coral Garden, Japanese Shipwreck and Tulamben. The overall average in Amed for percentage of cover is 43 percent, which falls in the medium category. Respondents suggested that Amed's reefs are healthy and improving. "I bought a dive center here after working all over Southeast Asia," said one respondent. "The reefs are healthy, and I consider this the last remaining paradise in Bali." Other respondents agreed that the reefs of Amed were healthy compared to other reefs on the

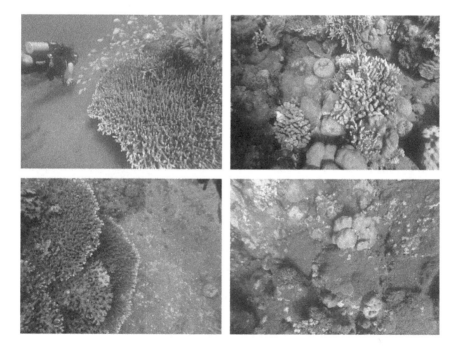

Figure C.4 Survey images from Amed, Indonesia. Top left: A divemaster leading tourists in Jemeluk Bay. Top right: Survey in Coral Garden. Bottom left: Survey in Japanese Shipwreck. Bottom right: Survey in Tulamben.

island and that they were improving. Many respondents cited concerns with bleaching and climate change.

C.3.4 Perhentian Islands, Malaysia

The images in Figure C.5 are from survey sites on Batu Layar, Shark Point, Tokong Laut and Achin. The overall average in the Perhentians for percentage of cover is 31.6 percent, which falls in the medium category. Respondents suggested that the reefs of the Perhentians, while still able to draw many tourists, were quickly becomming less healthy. Many cited increasingly non-existent enforcement of fishing bans by Marine Parks, as well as an overall lack of landward planning for both solid waste removal and erosion. "Water quality will kill the reefs," said a bar owner on the islands, "but we have nowhere to put the sewage, and every year more people clear jungle to build bars and restaurants." Many stakeholders cited repeat years of noticeably bad bleaching as additional reasons why they feared for the survival of the local reef.

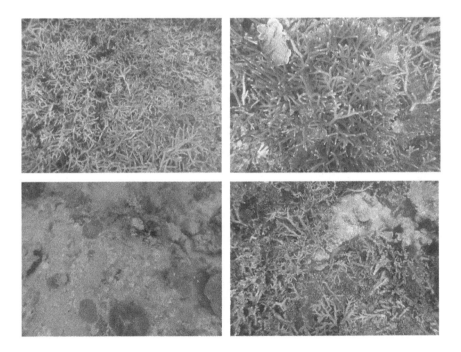

Figure C.5 Survey images from Perhentian Islands, Malaysia. Top left: Batu Layar. Top right: Shark Point. Bottom left: Tokong Luat. Bottom right: Achin.

Figure C.6 Survey images from Tioman Island, Malaysia. Top left: Tekek House Reef. Top right: Labas. Bottom: Renggis North.

C.3.5 Tioman Island, Malaysia

The images in Figure C.6 are from survey sites on House Reef, Labas and Renggis. The overall average on Tioman for percentage of cover is 56.25 percent, which falls in the good category. Respondents suggested that Tioman's reefs are rapidly declining thanks to damage from anchors, in addition to a lack of enforcement of rules in the park by some dive operators and Marine Parks. "People don't know what they have," said a hotel worker. "When I was a kid the fish were huge. Now we can't see anything out there. There's no fish. The coral looks good for now, but the fish are gone." Respondents also repeatedly mentioned climate change as a major concern.

REFERENCES

Abdul, J. (1999). An introduction to Pulau Tioman. *The Raffles Bulletin of Zoology*. National University of Singapore. Supplement No. 6, 3–4.

Adger, W. N. (1999). Social vulnerability to climate change and extremes in coastal Vietnam. *World Development*, 27(2), 249–269.

Adger, W. N., Brown, K., Fairbrass, J., Jordan, A., Paavola, J., Rosendo, S., and Seyfang, G. (2003). Governance for sustainability: towards a 'thick' analysis of environmental decisionmaking. *Environment and Planning A*, 35(6), 1095–1110.

Adger, W. N., Hughes, T. P., Folke, C., Carpenter, S. R., and Rockström, J. (2005). Social-ecological resilience to coastal disasters. *Science*, 309(5737), 1036–1039.

Acheson, J. M., Wilson, J. A., and Steneck, R. S. (1998). Managing chaotic fisheries. In F. Berkes and C. Folke (Eds.), *Linking social and ecological systems: Management practices and social mechanisms for building resilience* (pp. 390–413). Cambridge: Cambridge University Press.

Anderies, J. M., Janssen, M., and Ostrom, E. (2004). A framework to analyze the robustness of social-ecological systems from an institutional perspective. *Ecology and Society* 9(1), 18.

Agrawal, A., and Gibson, C. G. (1999). Enchantment and Disenchantment: The Role of Community in Natural Resource Conservation. *World Development* 27, 629–649.

Agrawal, A. (2001). Common property institutions and sustainable governance of resources. *World Development*, 29(10), 1649–1672.

Aichi Biodiversity Targets. (2016). Retrieved February 04, 2016, from https://www.cbd.int/sp/targets/

Alcorn, J. B. (1993). Indigenous peoples and conservation. *Conservation Biology*, 7(2), 424–426.

Alm, J., and Bahl, R. (1999). Decentralization in Indonesia: Prospect and problems. PEG Report No. 12, ECG, USAID/Jakarta. Atlanta: Department of Economics, the School of Policy Studies, Georgia State University.

Ananta, A., Soekarni, M., and Arifin, S. (Eds.). (2011). *The Indonesian economy: Entering a new era*. Singapore: Institute of Southeast Asian Studies.

Andaya, B. W., and Andaya, L. Y. (2001). *A history of Malaysia*. Honolulu: University of Hawaii Press.

Armitage, D. (2005). Adaptive capacity and community-based natural resource management. *Environmental Management*, 35(6), 703–715.

Armitage, D., Berkes, F., and Doubleday, N. (Eds.). (2007). *Adaptive co-management: collaboration, learning and multi-level governance*. Vancouver, Canada: University of British Columbia Press.

Badruddin, M., and Gillett, R. (1996). Translations of Indonesian fisheries law relevant to fisheries management in the extended economic zone. Unpublished report, FAO project: Strengthening marine fisheries development in Indonesia, Technical paper, 9.

Bali Government Tourism Office. 2015. Retrieved from http://www.disparda.baliprov.go.id/en/Statistics2

Becker, C. D., and Ostrom, E. (1995). Human ecology and resource sustainability: the importance of institutional diversity. *Annual Review of Ecology and Systematics*, 26, 113–133.

Bellwood, D., Hughes, T., Folke, C., and Nyström, M. (2004). Confronting the coral reef crisis. *Nature*, 429(6994), 827–833.

Berkes, F. (2007). Community-based conservation in a globalized world. *Proceedings of the National Academy of Sciences*, 104(39), 15188–15193.

Berkes, F. (2011). Co-management and the James Bay Agreement. In E. Pinkerton (Ed.), *Co-operative management of local fisheries: New directions for improved management and community development* (p. 189). Vancouver, Canada: University of British Columbia Press.

Berkes, F., George, P. J., and Preston, R. J. (1991). *Co-management: The evolution of the theory and practice of joint administration of living resources* (pp. 12–18). Program for Technology Assessment in Subarctic Ontario, McMaster University.

Berkes, F., and Folke, C. (1998). Linking social and ecological systems for resilience and sustainability. In *Linking social and ecological systems: management practices and social mechanisms for building resilience* (pp. 13–20). Cambridge: Cambridge University Press.

Berkes, F., Colding, J., and Folke, C. (2003). *Navigating social-ecological systems: Building resilience for complexity and change*. Cambridge: Cambridge University Press.

Bonham, C. A., Sacayon, E., and Tzi, E. (2008). Protecting imperiled "paper parks": potential lessons from the Sierra Chinajá, Guatemala. *Biodiversity and Conservation*, 17(7), 1581–1593.

Bourchier, D., and Hadiz, V. (Eds.). (2014). *Indonesian politics and society: A reader*. New York: Routledge.

Boyatzis, R. E. (1998). *Transforming qualitative information: Thematic analysis and code development*. Thousand Oaks, CA: Sage.

Braun, V., and Clarke, V. (2006). Using thematic analysis in psychology. *Qualitative Research in Psychology*, 3(2), 77–101.

Brewer, J., and Hunter, A. (1989). *Multimethod research: A synthesis of styles*. Newbury Park, NJ: Sage.

Brooks, J., Franzen, M., Holmes, C., Grote, M., and Mulder, M. (2006). Testing hypotheses for the success of different conservation strategies. *Conservation Biology*, 20(5), 1528–1538.

Brosius, J. P., Tsing, A. L., and Zerner, C. (1998). Representing communities: Histories and politics of community-based natural resource management. *Society & Natural Resources*, 11(2), 157–168.

Brown, B. (1997). Coral bleaching: Causes and consequences. *Coral Reefs*, 16(1), S129–S138.

Brown, C. (2003). *A short history of Indonesia: The unlikely nation?* Crown Nest, New South Wales: Allen and Unwin.

Buchanan, A., and Keohane, R. O. (2006). The legitimacy of global governance institutions. *Ethics and International Affairs*, 20(4), 405–437.

Burke, L., Selig, L., and Spalding, M. (2002). Reefs at risk in Southeast Asia. World Resources Institute.

Bythell, J. C., Hillis-Starr, Z. M., and Rogers, C. S. (2000). Local variability but landscape stability in coral reef communities following repeated hurricane impacts. *Marine Ecology Progress Series*, 204, 93–100.

Camacho, A. E., Susskind, L. E., and Schenk, T. (2010). Collaborative planning and adaptive management in Glen Canyon: A cautionary tale. *Columbia Journal of Environmental Law*, 35(1).

CBD Convention on Biolodigcal Diversity (CBD). (2016a). Indonesia - Country Profile. Retrieved February 04, 2016, from https://www.cbd.int/countries/profile/default. shtml?country=id#measures

CBD Convention on Biolodigcal Diversity (2016b). Malaysia - Country Profile Retrieved February 04, 2016, from https://www.cbd.int/countries/profile/default.shtml? country=my

Christie, P. (2004). Marine protected areas as biological successes and social failures in southeast asia. Paper presented at the American Fisheries Society Symposium 42 (pp. 155–164).

Christie, P., and White, A. T. (1997). Trends in development of coastal area management in tropical countries: from central to community orientation. *Coastal Management*, 25, 155–181

Christie, P., Bradford, D., Garth, R., Gonzalez, B., Hostetler, M., Morales, O., ... and White, N. (2000). *Taking care of what we have: Participatory natural resource management on the Atlantic Coast of Nicaragua*. University of Central America, International Development Research Centre, Managua, Ottawa.

Christie, P., White, A., and Deguit, E. (2002). Starting point or solution? Community-based marine protected areas in the Philippines. *Journal of Environmental Management*, 66(4), 441–454.

Christie, P., Buhat, D., Garces, L. R. and White, A. T. (2003). The challenges and rewards of community-based coastal resources management: San Salvador Island Philippines. In S. R. Brechin, P. R. Wilshusen, P. R. Fortwangler, and P. C. West (Eds.), *Contested nature—promoting international biodiversity conservation with social justice in the twenty-first century* (pp. 231–249). New York: SUNY Press.

Christie, P., and White, A. T. (2007). Best practices for improved governance of coral reef marine protected areas. *Coral Reefs*, 26(4), 1047–1056.

Cinner, J., and Huchery, C. (2014). A comparison of social outcomes associated with different fisheries co-management institutions. *Conservation Letters*, 7(3), 224–232.

Cinner, J. E., McClanahan, T. R., Daw, T. M., Graham, N. A., Maina, J., Wilson, S. K., and Hughes, T. P. (2009a). Linking social and ecological systems to sustain coral reef fisheries. *Current Biology*, 19(3), 206–212.

Cinner, J. E., Wamukota, A., Randriamahazo, H., and Rabearisoa, A. (2009b). Toward institutions for community-based management of inshore marine resources in the Western Indian Ocean. *Marine Policy*, 33(3), 489–496.

Cinner, J. E., McClanahan, T. R., MacNeil, M. A., Graham, N. A., Daw, T. M., Mukminin, Kuange, J. (2012). Comanagement of coral reef social-ecological systems. *Proceedings of the National Academy of Sciences*, 109(14), 5219–5222.

Colfer, C. J. P., and Capistrano, D. (2005). *The politics of decentralization: Forests, people and power*. London: Earthscan.

Comoros: Coastal Resources Co-management for Sustainable Livelihood. (2015, December 22). Retrieved February 04, 2016, from http://www.worldbank.org/projects/ P125301/comoros-coastal-resources-co-management-sustainable-livelihood?lang=en

Corbin, J. M., and Strauss, A. (1990). Grounded theory research: Procedures, canons, and evaluative criteria. *Qualitative Sociology*, 13(1), 3–21.

Corbin, J., and Strauss, A. (1996). Analytic ordering for theoretical purposes. *Qualitative Inquiry*, 2(2), 139–150.

COREMAP. (2007). Coremap 2014. Retrieved from http://www.coremap.or.id/tentang_ coremap/mengenal_coremap/

Costanza, R., d'Arge, R., de Groot, R., Farber, S., Grasso, M., Hannon, B., et al. (1997). The value of the world's ecosystem services and natural capital. *Nature*, 387(6630), 253–260.

CTI-Coral Triangle Institute (2014). Indonesia's COREMAP-CTI Gets New US$47.38M Loan from World Bank. http://www.coraltriangleinitiative.org/news/indonesia%E2%80%99s-coremap-cti-gets-new-us4738m-loan-world-bank

Dahuri, R., and Dutton, I. M. (2000). Integrated coastal and marine management enters a new era in Indonesia. *Integrated Coastal Zone Management*, 1, 11–16.

Daniels, S. E., and Walker, G. B. (1996). Collaborative learning: improving public deliberation in ecosystem-based management. *Environmental Impact Assessment Review*, 16(2), 71–102.

de Oliveira, J. A. P. (2002). Implementing environmental policies in developing countries through decentralization: the case of protected areas in Bahia, Brazil. *World Development*, 30(10), 1713–1736.

Department of Marine Parks Malaysia. (2014). Retrieved from http://www.dmpm.nre.gov.my/data_pelawat.html?uweb=jtlandlang=en

de Groot, R. S., Wilson, M. A., and Boumans, R. M. (2002). A typology for the classification, description and valuation of ecosystem functions, goods and services. *Ecological Economics*, 41(3), 393–408.

de Oliveira, J. A. P. (2002). Implementing environmental policies in developing countries through decentralization: the case of protected areas in Bahia, Brazil. *World Development*, 30(10), 1713–1736.

de Vries, I. (2011). Indonesia's infrastructure problems: A legacy from Dutch colonialism. *Jakarta Globe*. Retrieved February 06, 2016, from http://jakartaglobe.beritasatu.com/archive/indonesias-infrastructure-problems-a-legacy-from-dutch-colonialism/

Dietz, T., Ostrom, E., and Stern, P. C. (2003). The struggle to govern the commons. *Science*, 302(5652), 1907–1912.

Dirhamsyah. (2005). Analysis of the effectiveness of Indonesia's coral reef management framework. PhD thesis, Centre for Maritime Policy, University of Wollongong, Australia.

Dodson, G. (2014). Co-governance and local empowerment? Conservation partnership frameworks and marine protection at Mimiwhangata, New Zealand. *Society and Natural Resources*, 27(5), 521–539.

Doucet, A., and Mauthner, N. (2008). Qualitative interviewing and feminist research. In P. Alasuutari, L. Bickman, and J. Brannen (Eds.), *The SAGE handbook of social research methods*. London: Sage.

Dunning, K. H. (2015). Ecosystem services and community based coral reef management institutions in post blast-fishing Indonesia. *Ecosystem Services*, 16, 319–332.

Dutton, I. M. 2005. If only fish could vote: The enduring challenges of coastal and marine resources management in post-Reformasi Indonesia. In B. P. Resosudarmo (Ed.), *The politics and economics of Indonesia's Natural Resources* (pp. 162–178). Singapore: Institute of Southeast Asian Studies.

Elliott, G., Mitchell, B., Wiltshire, B., Manan, I. A., and Wismer, S. (2001). Community participation in marine protected area management: Wakatobi National Park, Sulawesi, Indonesia. *Coastal Management*, 29(4), 295–316.

English, S. S., Wilkinson, C. C., and Baker, V. V. (1994). *Survey manual for tropical marine resources*. Townsville, Australian Institute of Marine Science.

Erdmann, M. (2001). Who's minding the reef? Corruption and enforcement in Indonesia. *SPC Live Reef Fish Information Bulletin*, 8, 19–20.

Erviani. (2009). Bali wants more in tourism revenue. *Jakarta Post*. Retrieved from http://www.thejakartapost.com/news/2009/04/08/bali-wants-more-tourism-revenue.html-0

Flyvbjerg, B. (2006). Five misunderstandings about case-study research. *Qualitative Inquiry*, 12(2), 219–245.

Folke, C., Colding, J., and Berkes, F. (2003). Synthesis: Building resilience and adaptive capacity in socio-ecological systems. In F. Berkes, C. Folke, and J. Colding (Eds.), *Navigating social–ecological systems: Building resilience for complexity and change* (pp. 352–387). Cambridge: Cambridge University Press.

Glover, L., and Earle, S. (Eds.). (2004). *Defying ocean's end: An agenda for action*. Washington, DC: Island Press.

Glynn, P. W. (1993). Coral reef bleaching: Ecological perspectives. *Coral Reefs*, 12(1), 1–17.

Goreau, T., McClanahan, T., Hayes, R., and Strong, A. L. (2000). Conservation of coral reefs after the 1998 global bleaching event. *Conservation Biology*, 14(1), 5–15.

Gomez, E. D., and Yap, H. T. (1988). Monitoring reef condition. Coral Reef Management Handbook. UNESCO Regional Office for Science and Technology for Southeast Asia (ROSTSEA). Jakarta.

Gunderson, L. (1999). Resilience, flexibility and adaptive management—antidotes for spurious certitude. *Conservation Ecology*, 3(1), 7.

Gunderson, L. H. and Holling, C. S. (Eds.). (2001). Panarchy: *Understanding transformations in human and natural systems*. Washington, DC: Island Press.

Abdullah, H., Setiasih, N., and Sartin, J. (2007). A decade of Reef Check monitoring: Indonesian coral reefs, condition and trends. *The Indonesian Reef Check Network*, 32.

Hadiz, V. R. (2004). Decentralization and democracy in Indonesia: A critique of neo-institutionalist perspectives. *Development and Change*, 35(4), 697–718.

Hanson, W., Creswell, J., Clark, V., Petska, K., and Creswell, J. (2005). Mixed methods research designs in counseling psychology. *Journal of Counseling Psychology*, 52(2), 224.

Harborne, A., Fenner, D., Bames, A., Beger, M., Harding, S., and Roxburgh, T. (2000). Status report on coral reefs of the East Coast of Peninsular Malaysia, Kuala Lumpur. Report prepared to Department of Fisheries, Malaysia. Retrieved from http://www.dmpm.nre.gov.my/files/Status%20Report%20on%20The%20Reefs%20of%20The%20East%20Coast%20of%20Peninsular%20Malaysia.pdf

Harborne, A., Fenner, D., Barnes, A., Beger, M., Harding, S., and Roxburgh, T. (2000). Status report on the coral reefs of the east coast of Peninsula Malaysia. Report Prepared to Department of Fisheries Malaysia, Kuala Lumpur, Malaysia, 361–69.

Harrell, M. C., and Bradley, M. A. (2009). Data collection methods. Semi-structured interviews and focus groups. Rand National Defense Research Institute. Santa Monica, CA.

Harvell, C. D., Mitchell, C. E., Ward, J. R., Altizer, S., Dobson, A. P., Ostfeld, R. S., and Samuel, M. D. (2002). Climate warming and disease risks for terrestrial and marine biota. *Science*, 296(5576), 2158–2162.

Hawkins, J. P., Roberts, C. M., Kooistra, D., Buchan, K., and White, S. (2005). Sustainability of scuba diving tourism on coral reefs of Saba. *Coastal Management*, 33(4), 373–387.

Hayes, N. (1997). Theory-led thematic analysis: social identification in small companies. In N. Hayes (Ed.), *Doing qualitative analysis in psychology*. Hove, UK: Psychology Press.

Henley, D. (2008). Natural resource management: Historical lessons from Indonesia. *Human Ecology*, 36(2), 273–290.

Hughes, T. P., Baird, A. H., Bellwood, D. R., Card, M., Connolly, S. R., Folke, C., Grosberg, R., Hoegh-Guldberg, O., Jackson, J. B. C., Kleypas, J., Lough, J. M., Marshall, P., Nystrom, M., Palumbi, S. R., Pandolfi, J. M., Rosen, B., and Roughgarden, J. (2003). Climate change, human impacts, and the resilience of coral reefs. *Science*, 301:929–933.

Hamilton, R. J., Potuku, T., and Montambault, J. R. (2011). Community-based conservation results in the recovery of reef fish spawning aggregations in the Coral Triangle. *Biological Conservation*, 144(6), 1850–1858.

Hill, J., and Wilkinson, C. (2004). *Methods for ecological monitoring of coral reefs*. Townsville: Australian Institute of Marine Science, 117.

Holling, C. S. (1973). Resilience and stability of ecological systems. *Annual Review of Ecology and Systematics*, 1–23.

Holling, C. S. (1978). *Adaptive environmental assessment and management*. Chichester, UK: Wiley-Interscience.

Hughes, T. P., Bellwood, D. R., Folke, C., Steneck, R. S., and Wilson, J. (2005). New paradigms for supporting the resilience of marine ecosystems. *Trends in Ecology and Evolution*, 20(7), 380–386.

ICRI International Coral Reef Initiative. (2010). Indonesia - Global Mass Bleaching of Coral Reefs in 2010. Retrieved from http://www.icriforum.org/news/2010/08/indonesia-global-mass-bleaching-coral-reefs-2010

ICRI International Coral Reef Institute. (2010). Malaysia Marine Protected Areas. Retrieved from http://earw.icriforum.org/2010/11.Malaysia_(Irwan_Isnain).pdf

Imperial, M. T., and Yandle, T. (2005). Taking institutions seriously: Using the IAD framework to analyze fisheries policy. *Society and Natural Resources*, 18(6), 493–509.

Ingold, T. (2000). *The perception of the environment: essays on livelihood, dwelling and skill*. Hove, UK: Psychology Press.

Islam, G. M. N., Noh, K. M., Yew, T. S., and Noh, A. F. M. (2013). Assessing environmental damage to marine protected area: a case of Perhentian Marine Park in Malaysia. *Journal of Agricultural Science*, 5(8), 132.

Islam, G., Yew, T., Noh, K., Noh, A. (2014). Community's perspectives towards Marine Protected Area in Perhentian Marine Park Malaysia. *Open Journal of Marine Science*, 4, 51–60.

IUCN International Union for the Conservation of Nature. The World Conservation Union (1988) Resolution 17.38 of the 17th session of the General Assembly of the IUCN. Gland, Switzerland.

IUCN International Union for the Conservation of Nature. The World Conservation Union (1994) Resolution 19.46 of the 19th session of the General Assembly of the IUCN, Buenos Aires, Argentina.

Jackson, J. B., Kirby, M. X., Berger, W. H., Bjorndal, K. A., Botsford, L. W., Bourque, B. J., and Hughes, T. P. (2001). Historical overfishing and the recent collapse of coastal ecosystems. *Science*, 293(5530), 629–637.

Jentoft, S., McCay, B. J., and Wilson, D. C. (1998). Social theory and fisheries co-management. *Marine Policy*, 22(4), 423–436.

Jick, T. D. (1979). Mixing qualitative and quantitative methods: Triangulation in action. *Administrative Science Quarterly*, 602–611.

Johannes, R. E. (1981). *Words of the lagoon: Wshing and marine lore in the Palau District of Micronesia*. Berkeley: University of California Press.

Jones, E. V., Gray, T., Macintosh, D., and Stead, S. (2016). A comparative analysis of three marine governance systems for implementing the Convention on Biological Diversity (CBD). *Marine Policy*, 66, 30–38.

Juda, L. (1999). Considerations in the development of a functional approach to the governance of large marine ecosystems. *Ocean Development & International Law*, 30, 89–125.

Jul-Larsen, E., Kolding, J., Overå, R., Nielsen, J., and Van Zwieten, P. A. (2003). Management, co-management or no management? Major Dilemmas in Southern African Freshwater Fisheries 1. Synthesis Report. Retrieved February 04, 2016, from http://www.fao.org/docrep/005/y4593e/y4593e00.HTM

Kahn, J. S. (Ed.). (1998). *Southeast Asian identities: Culture and the politics of representation in Indonesia, Malaysia, Singapore, and Thailand*. New York: St. Martin's Press.

Kennedy, C. (2015, March 3) Warm oceans pose risk of global coral bleaching event in 2015. Retrieved from https://www.climate.gov/news-features/featured-images/warm-oceans-pose-risk-global-coral-bleaching-event-2015

Khai L. H. (1992). Dynamics of policy-making in Malaysia: The formulation of the New Economic Policy and the National Development Policy. *Asian Journal of Public Administration*, 14(2), 204–227.

King, G., Keohane, R. O., and Verba, S. (1994). *Designing social inquiry: Scientific inference in qualitative research*. Princeton, NJ: Princeton University Press.

Kittinger, J. N. (2013). Participatory fishing community assessments to support coral reef fisheries comanagement 1. *Pacific Science*, 67(3), 361–381.

Koontz, T. M., and Johnson, E. M. (2004). One size does not fit all: Matching breadth of stakeholder participation to watershed group accomplishments. *Policy Sciences*, 37(2), 185–204.

Kothari, U. (2001). Power, knowledge and social control in participatory development. In B. Cooke and U. Kothari, (Eds.), *Participation: the New Tyranny?* (pp 139–152). London: Zed Books.

Laman Web Rasmi Taman Laut Malaysia. (2016). 2010 List of Closed Areas. Retrieved from http://www.dmpm.nre.gov.my/senarai_kawasan_yang_ditutup.html?uweb=jtl

Laman Web Rasmi Jabatan Taman Laut Malaysia. (2016, January 29). Retrieved February 5, 2016, from http://www.dmpm.nre.gov.my/index.php?andlang=en

Leach, M., Mearns, R., and Scoones, I. (1999). Environmental entitlements: Dynamics and institutions in community-based natural resource management. *World Development*, 27(2), 225–247.

Lee, K. N. (1994). *Compass and gyroscope: Integrating science and politics for the environment*. Washington, DC: Island Press.

Lipsky, M., and Hurley, S. (2009). Design Sensitivity. In L. Bickman and D. J. Rog (Eds.), *The SAGE handbook of applied social research methods*. Los Angeles, CA: Sage.

March, J. G., and Olsen, J. P. (1989). *Rediscovering institutions*. New York: Simon & Schuster.

Matheson-Hooker, V. (2003). *A short history of Malaysia: Linking East and West*. Crows Nest, Australia: Allen and Unwin.

McClanahan, T. R. (2000). Recovery of a coral reef keystone predator, Balistapus undulatus, in East African marine parks. *Biological Conservation*, 94:191–198.

McClanahan, T. R., Castilla, J. C., White, A., and Defeo, O. (2006). Healing small-scale fisheries and enhancing ecological benefits by facilitating complex social–ecological systems. *Reviews in Fish Biology and Fisheries 2009*, 19:33–47.

MEA Millennium ecosystem assessment. (2003). *Eco systems and human well-being: A framework for assessment.* Washington, DC: Island Press.

Merrill, R. (1998). The NRMP Experience in Bunaken and Bukit Baka-Bukit Raya National Parks: Lessons Learned for PAM in Indonesia, National Resources Management Project, Jakarta.

Ministry of Marine Affairs and Fisheries. (2003). Urgensi RUU Pengelolaan Wilayah Pesisir dan Pulau-Pulau Kecil. Retrieved from https://uwityangyoyo.wordpress.com/2009/11/12/pengelolaan-wilayah-pesisir-secara-terpadu-dan-berkelanjutan-yang-berbasis-masyarakat/.

Moberg, F., and Folke, C. (1999). Ecological goods and services of coral reef ecosystems. *Ecological Economics*, 29(2), 215–233.

Mora, C., Andréfouët, S., Costello, M. J., Kranenburg, C., Rollo, A., Veron, J., … and Myers, R. A. (2006). Coral reefs and the global network of marine protected areas. *Science*, 312:1750–1751

Moran, E. F. (2010). *Environmental social science*: Human-environment interactions and sustainability (pp. i-xiii). Malden, MA: Wiley-Blackwell.

Moran, E. F., and Ostrom, E. (2005). *Seeing the Forest and the Trees: Human-environment interactions in forest ecosystems.* Cambridge, MA: Mit Press.

Mulrennan, M. E., Mark, R., and Scott, C. H. (2012). Revamping community-based conservation through participatory research. *The Canadian Geographer/Le Géographe Canadien*, 56(2), 243–259.

Mustika, P. L. K. (2011). Towards sustainable dolphin watching tourism in Lovina, Bali, Indonesia. PhD dissertation, James Cook University.

Mytelka, L. K., and Smith, K. (2002). Policy learning and innovation theory: An interactive and co-evolving process. *Research Policy*, 31(8), 1467–1479.

NRC National Research Council (US). Committee on the Evaluation, Design, and Monitoring of Marine Reserves and Protected Areas in the United States. (1999). Marine protected areas: Tools for sustaining ocean ecosystems. National Academy Press.

OECD. (2014). Tourism Trends. OECD. http://www.mlit.go.jp/kankocho/naratourism statisticsweek/statistical/pdf/2014_OECD_Tourism_Trends.pdf

Olsson, P., C. Folke, and T. Hahn. 2004. Socio-ecological transformation for ecosystem management: The development of adaptive co-management of a wetland landscape in southern Sweden. *Ecology and Society*, 9(4):2. Retrieved from http://www.ecologyandsociety.org/vol9/iss4/art2.

Ostrom, E. (1990). *Governing the commons: The evolution of institutions for collective action.* Cambridge: Cambridge University Press.

Ostrom, E. (2007). A diagnostic approach for going beyond panaceas. *Proceedings of the national Academy of Sciences*, 104(39), 15181–15187.

Ostrom, E. (2008). The challenge of common-pool resources. *Environment: Science and Policy for Sustainable Development*, 50(4), 8–21.

Ostrom, E. (2009). *Understanding institutional diversity.* Princeton, NJ: Princeton University Press.

Pandolfi, J. M., Bradbury, R. H., Sala, E., Hughes, T. P., Bjorndal, K. A., Cooke, R. G., … and Warner, R. R. (2003). Global trajectories of the long-term decline of coral reef ecosystems. *Science*, 301(5635), 955–958.

Peet, R., and Watts, M. (2004). *Liberation ecologies: Environment, development, social movements.* Hove, UK: Psychology Press.

Peluso, N. L. (1992). *Rich forests, poor people: Resource control and resistance in Java*. Berkeley: University of California Press.

Pet-Soede, C., Cesar, H. S. J., and Pet, J. S. (1999). An economic analysis of blast fishing on Indonesian coral reefs. *Environmental Conservation*, 26(02), 83–93.

Peterson, C. H., Lubchenco, J., and Daily, G. (1997). *Marine ecosystem services*. Washington, DC: Island Press.

Pinkerton, E. (Ed). (1989). *Co-operative management of local fisheries new directions for improved management and community development*. Vancouver, Canada: University of British Columbia Press.

Pinkerton, E. W. (1994). Local fisheries co-management: A review of international experiences and their implications for salmon management in British Columbia. *Canadian Journal of Fisheries and Aquatic Sciences*, 51(10), 2363–2378.

Plummer, R., and Fitzgibbon, J. (2004). Co-management of natural resources: A proposed framework. *Environmental Management*, 33(6), 876–885.

Pollnac, R. B., and Pomeroy, R. S. (2005). Factors influencing the sustainability of integrated coastal management projects in the Philippines and Indonesia. *Ocean and Coastal Management*, 48(3), 233–251.

Pomeroy, R. S. (1995). Community-based and co-management institutions for sustainable coastal fisheries management in Southeast Asia. *Ocean and Coastal Management*, 27(3), 143–162.

Pomeroy, R., and Douvere, F. (2008). The engagement of stakeholders in the marine spatial planning process. *Marine Policy*, 32(5), 816–822.

Rakyat Post. (2014, June 4). Coral Bleaching May Cause Malaysian Dive Sites to Be Closed. Retrieved from http://www.therakyatpost.com/news/2014/06/04/coral-bleaching-may-cause-malaysian-dive-sites-closed/

Rakyat Post. (2016, July 7). Sabah Parks Director disappointed over unethical behavior of some dive operators. Retrieved from http://www.therakyatpost.com/news/2015/07/07/sabah-parks-director-disappointed-over-unethical-behaviour-of-some-dive-operators/

Rakyat Post. (2015, August 26). What you need to know about Malaysia's Bersih movement. Retrieved from http://www.straitstimes.com/asia/se-asia/what-you-need-to-know-about-malaysias-bersih-movement

Reed, M. S. (2008). Stakeholder participation for environmental management: a literature review. *Biological Conservation*, 141(10), 2417–2431.

Reed, M. S., Dougill, A. J., and Baker, T. R. (2008). Participatory indicator development: what can ecologists and local communities learn from each other. *Ecological Applications*, 18(5), 1253–1269.

Reef Check Malaysia. (2010). Annual Survey Report. Reef Check Malaysia.

Reef Check News. (2009, June 29) Up to 40% Bleaching Recorded in Bali. Retrieved from http://www.reefcheck.org/reef-news/up-to-40-coral-bleaching-recorded-in-bali

Renard, Y. (1991). Institutional challenges for community-based management in the caribbean. *Nature and Resources*, 74(4), 4–9.

Renn, O., Webler, T., Wiedemann, P. (Eds.). (1995). *Fairness and competence in citizen participation*. Dordrecht: Kluwer Academic Publishers.

Richards, C., Blackstock, K. L., and Carter, C. E., 2004. Practical approaches to participation SERG Policy Brief No. 1. Macauley Land Use Research Institute, Aberdeen.

Ribot, J. (2002). *Democratic decentralization of natural resources: Institutionalizing popular participation.* Washington, DC: World Resources Institute.

Richmond, R. H. (1993). Coral reefs: Present problems and future concerns resulting from anthropogenic disturbance. *American Zoologist*, 33(6), 524–536.

Rogers, C. S., Garrison, G., Grober, R., Hillis, Z. M., and Franke, M. A. (1994). *Coral reef monitoring manual for the Caribbean and Western Atlantic.* St. John, US Virgin Islands: National Park Service.

Rowe, G., Frewer, L., 2000. Public participation methods: A framework for evaluation in science. *Technology and Human Values*, 25, 3–29.

Ruddle, K. (1988). Social principles underlying traditional inshore fishery management systems in the Pacific Basin. *Marine Resource Economics*, 5, 351–363.

Ruddle, K. (1994). *A guide to the literature on traditional community-based fishery management in the Asia-Pacific Tropics.* FAO Fisheries Circular 869, Rome, Italy.

Saad, J. (2013). *Review of Malaysian laws and policies in relation to the implementation of the ecosystem approach to fisheries management in Malaysia.* Honolulu, Hawaii: The USAID Coral Triangle Support Partnership.

Salvat, B. (1992). Coral reefs—a challenging ecosystem for human societies. *Global Environmental Change*, 2, 12–18.

Sandersen, H. T., and Koester, S. (2000). Co-management of tropical coastal zones: The case of the Soufriere marine management area, St. Lucia, WI. *Coastal Management*, 28(1), 87–97.

Savitri, A. (2001, March 8). Pemuteran village successful in boosting tourism. *Jakarta Post.*

Scott, J. C. (1998). *Seeing like a state: How certain schemes to improve the human condition have failed.* New Haven, CT: Yale University Press.

Sheppard, C. R., Davy, S. K., and Pilling, G. M. (2009). *The biology of coral reefs.* Oxford: Oxford University Press.

Stevenson, T. C., and Tissot, B. N. (2014). Current trends in the analysis of co-management arrangements in coral reef ecosystems: a social–ecological systems perspective. *Current Opinion in Environmental Sustainability*, 7, 134–139.

Suman, D., Shivlani, M., and Milon, J. W. (1999). Perceptions and attitudes regarding marine reserves: a comparison of stakeholder groups in the Florida Keys National Marine Sanctuary. *Ocean and Coastal Management*, 42(12), 1019–1040.

Susskind, L., Camacho, A. E., and Schenk, T. (2012). A critical assessment of collaborative adaptive management in practice. *Journal of Applied Ecology*, 49(1), 47–51.

Tan, C. H., and Heron, S. F. (2011). First observed severe mass bleaching in Malaysia, Greater Coral Triangle. *Galaxea*, 13(1), 27–28.

Tendler, J. (1997). *Good governance in the tropics.* Baltimore, MD: John Hopkins University Press.

Thorburn, C. (2002). Regime change—prospects for community-based resource management in post-New Order Indonesia. *Society and Natural Resources*, 15(7), 617–628.

Toda, T., Okashita, T., Maekawa, T., Alfian, B. A. A. K., Rajuddin, M. K. M., Nakajima, R., ... and Terazaki, M. (2007). Community structures of coral reefs around Peninsular Malaysia. *Journal of Oceanography*, 63(1), 113–123.

Tomascik, T., Mah, A. J., Nonji, A., and Moosa, M. K. (1997). *The ecology of the Indonesian seas Part One and Two.* Singapore: EMDI.

Tratalos, J. A., and Austin, T. J. (2001). Impacts of recreational SCUBA diving on coral communities of the Caribbean island of Grand Cayman. *Biological Conservation*, 102(1), 67–75.

UNDP United Nations Development Programme. (2013). Conserving marine biodiversity through enhanced marine park management and inclusive island development. Retrieved from http://www.undp.org/content/dam/malaysia/docs/EnE/EnEProDocs/Conserving%20Marine%20Biodiversity%20through%20Enhanced%20Marine%20Park%20Management%20and%20Inclusive%20Sustainable%20Island%20Development%20Prodoc.pdf

UNEP United Nations Environmental Programme. (2005). Indonesia: Integrated assessment of the poverty reduction strategy paper with a case study on sustainable fisheries.

Uphoff, N. (1991). Fitting projects to people. In M. M. Cemea (Ed.), *Putting people first: Sociological variables in rural development*. Published for the World Bank. Oxford: Oxford University Press.

UP-MSI, ABC, ARCBC, DENR, ASEAN. (2002). Marine Protected Areas in Southeast Asia. ASEAN Regional Centre for Biodiversity Conservation, Department of Environment and Natural Resources, Los Baños, Philippines. 142 pp., 10 maps.

UNDP United Nations Development Programme. (2011). Conserving marine biodiversity through enhanced marine park management and sustainable island development. Mid Term Review. Retrieved March 1, 2015 from http://www.dmpm.nre.gov.my/files/Midterm%20Review%20Report.pdf

USAID/COMFISH. (2016). Collaborative management for a sustainable fisheries future in Senegal. Retrieved February 04, 2016, from http://www.crc.uri.edu/projects_page/senegalcomfish/

Veron, J. E. N., Devantier, L. M., Turak, E., Green, A. L., Kininmonth, S., Stafford-Smith, M., and Peterson, N. (2009). Delineating the coral triangle. *Galaxea*, 11(2), 91–100.

Veron, J. E. N., Hoegh-Guldberg, O., Lenton, T. M., Lough, J. M., Obura, D. O., Pearce-Kelly, P., ... and Rogers, A. D. (2009). The coral reef crisis: The critical importance of <350 ppm CO2. *Marine Pollution Bulletin*, 58(10), 1428–36.

Walker, B., Carpenter, S., Anderies, J., Abel, N., Cumming, G., Janssen, M., ... and Pritchard, R. (2002). Resilience management in socio-ecological systems: A working hypothesis for a participatory approach. *Conservation Ecology* 6(1), 14 [online]. Available at http://www.consecol.org/vol16/iss1/art14.

Walters, C. J. (1986). *Adaptive management of renewable resources*. New York: Macmillan; Collier Macmillan.

Walters, C. J., and Hilborn, R. (1978). Ecological optimization and adaptive management. *Annual Review of Ecology and Systematics*, 9, 157–88.

Wamukota, A. W., Cinner, J. E., and McClanahan, T. R. (2012). Co-management of coral reef fisheries: A critical evaluation of the literature. *Marine Policy*, 36(2), 481–488.

Wells, S. M. (1993). Coral reef conservation and management, progress in the South and Southeast Asian regions. *Coastal Management in Tropical Asia*, 1(1), 8–13.

West, Jordan M., and Rodney V. Salm. (2003). "Resistance and resilience to coral bleaching: implications for coral reef conservation and management." *Conservation Biology*, 17(4), 956–967.

White, A. T., Hale, L. Z., Renard, Y., and Cortesi, L. (1994). *Collaborative and community-based management of coral reefs: Lessons from experience*. Kumarian Press.

Wilkinson, C. R. (1998). The 1997–1998 mass bleaching event around the world. In C. R. Wilkinson (Ed.), *Status of coral reefs of the world: 1998* (pp. 15–38). Australian Institute of Marine Science, Cape Ferguson, Western Australia.

Wilkinson, C. (2008). Status of coral reefs of the world: 2008. Global Coral Reef Monitoring Network and Reef and Rainforest Research Centre, Townsville, Australia. No. 296.

World Bank. (2015a). Indonesia Data. http://data.worldbank.org/country/indonesia

World Bank. (2015b). Malaysia Data. http://data.worldbank.org/country/malaysia

World Bank. (2015c). Coral Reef Rehabilitation and Management Project II. Retrieved February 05, 2016, from http://www.worldbank.org/projects/P071318/coral-reef-rehabilitation-management-project-ii?lang=en

WTTC (World Tourism and Travel Council). (2015). Travel and Tourism Economic Impact 2015 Malaysia. Retrieved from https://www.wttc.org/-/media/files/reports/economic%20impact%20research/countries%202015/malaysia2015.pdf

Yewdall, K., Hammer, M., and Sticker, A. (2012). Paradise in Peril: Studying and protecting reefs, sharks, dolphins, and turtles of Pulau Tioman Marine Park.

Yahaya, J., and Yamamoto, T. (1988). A socioeconomic study of fisheries management and conservation with particular reference to two artisanal fishing villages in Penang, Peninsular Malaysia. College of Economics, Nihon University, CENU International Publications Series No. 1.

Yusof, Z. A., and Bhattasali, D. (2008). Economic growth and development in Malaysia: policy making and leadership. *International Bank for Reconstruction and Development/The World Bank.*

Zakai, D., and Chadwick-Furman, N. E. (2002). Impacts of intensive recreational diving on reef corals at Eilat, northern Red Sea. *Biological Conservation*, 105(2), 179–187.

Zerner, C. (2000). *People.* New York: Columbia University Press.

INDEX

Page numbers in *italics* and **bold** denote figures and tables, respectively.

INDEX 213

Lightning Source UK Ltd.
Milton Keynes UK
UKHW040222311018

331509UK00006B/244/P